Kinship and Human Evolution

Kinship and Human Evolution

Making Culture, Becoming Human

Steen Bergendorff

LEXINGTON BOOKS
Lanham • Boulder • New York • London

Published by Lexington Books
An imprint of The Rowman & Littlefield Publishing Group, Inc.
4501 Forbes Boulevard, Suite 200, Lanham, Maryland 20706
www.rowman.com

Unit A, Whitacre Mews, 26-34 Stannary Street, London SE11 4AB

British Library Cataloguing in Publication Information Available

Library of Congress Cataloging-in-Publication Data

Names: Bergendorff, Steen.
Title: Kinship and human evolution : making culture, becoming human / Steen Bergendorff.
Description: Lanham : Lexington Books, [2016] | Includes bibliographical references and index.
Identifiers: LCCN 2015042693 (print) | LCCN 2015046250 (ebook) | ISBN 9781498524179 (cloth : alk. paper) | ISBN 9781498524186 (electronic)
Subjects: LCSH: Kinship. | Human evolution.
Classification: LCC GN487 .B46 2016 (print) | LCC GN487 (ebook) | DDC 306.83—dc23 LC record available at http://lccn.loc.gov/2015042693

∞™ The paper used in this publication meets the minimum requirements of American National Standard for Information Sciences Permanence of Paper for Printed Library Materials, ANSI/NISO Z39.48-1992.

Printed in the United States of America

Contents

Introduction

Current explanations of human evolution are beset by problems. Notably the late transition of the human species into *Homo sapiens* is inadequately explained: no plausible adaptive reasons are offered for this evolution, or for how culture simultaneously arose. If we are to believe the literature, *Homo sapiens* seemingly evolved freely and not as a result of adaptation to the environment. This is truly exceptional. It is also impossible. If it were true, humans would be the only species ever to evolve in majestic isolation from the environment, driven only by its own innate potential.

We need an explanation for the transition into *Homo sapiens* that is rooted in adaptive theory, not in the myth of the Immaculate Conception. Humans, like any other contemporary species, arose as a result of natural selection, and natural selection always needs something to work with. In this book, I shall argue that that something was kinship. Kinship is arguably the most common human trait—everybody has it—and hence it must have played a major role in human evolution. The missing link in the story of human evolution is an adaptive explanation for the origins of kinship.

This book explains the adaptive role of kinship by looking at it as a survival strategy in a changing environment in which natural selection favored those who were able to make alliances. Having thus established that kinship was the major factor in the evolution of *Homo sapiens* it then becomes necessary to rethink our understanding of culture. Culture can no longer be seen as a characteristic of isolated groups that come into contact with one another. When kinship is seen as *the* major factor in human evolution this situation is turned around so that relationships precede cultural units. Networks and alliances then become the defining elements of localized culture.

There are several reasons why natural selection and adaption have been omitted, or ignored, in theories about the transition into *Homo sapiens* and the emergence of a culturally informed life. First, it has been disregarded because human evolution—in contrast to that of all other animals—has been seen as a cumulative process in which humans gradually improved their mental capabilities as a result of tool-making and increased sociability (e.g. Keesing 1981). This process also gradually made it possible for humans to colonize still wider environmental niches. And by colonizing these diverse niches they also became culturally different from one another. In this narrative, it seems that the human species became liberated from nature's grip by cumulatively enhancing their existing capabilities. This is in all probability an unlikely scenario, since it completely defies the rules of adaptation and selection.

Second, and linked to this, adaptation has been ignored in popular explanations that posit a major and irrevocable turning point from when prehistoric humans began their transformation into *sapiens*. No further adaptation was then presumed necessary. It has, for instance, been suggested that the domestication of fire (e.g., Burton 2009), or cooking (Wrangham 2009), or grooming (Dunbar 2004), respectively, were the major factors that triggered this transformation. Some even claim that it was the emergence of religious thought that did the trick (e.g., Wilson 2012)—only this already stipulates prior symbolic thought.

Even though all these abilities each vastly improved the living conditions of the early *Homo* species and still play a vital role in present human life, they were also evident among earlier *Homo* species, and they therefore cannot play a decisive role in the final transition into *Homo sapiens*. And however social pre-human groups became as a result, it is an entirely different matter to account for how and why these changes prompted humans to come up with their symbols and to begin to live in symbolically constituted communities. None of the aforementioned abilities involve any trace of symbolism.

Third—and perhaps the most significant reason why adaptation has been ignored in the study of human evolution—is because of a widespread idea that human development is the pinnacle of a general evolutionary process that is inevitable, progressive, and multiregional. Intelligence can, according to this view, in principle, evolve anywhere. As Tattersall writes, this is exactly what we all learned in school, where evolution "was usually presented as a matter of slow, inexorable refinement, constantly tending toward achieving the prefect" (2012: xv). Seemingly by lucky fortune, *Homo sapiens* was the end product of this process.

The incremental development of intelligence is believed to be such a universal process that we send space shuttles into space and search the sky for radio waves that could reveal the existence of other intelligent beings. It is assumed that there is a universal, evolutionary race toward intelligence, the only question being who will get there first. Science fiction is the ultimate example of this idea. Could some alien creature have won the competition already? Perhaps the *Star Trek* favorites—the Klingons, the Cardassians, or the Borgs? Or might some creature on Earth have won instead of us—monkeys, birds, or ants?

Science fiction captures the most popular view of evolution. But if we return to Earth we find two major, competing, scientific explanations of human evolution: the multiregional hypothesis and the Out-of-Africa hypothesis. Each has very different repercussions. The first hypothesis claims that evolution is a general process and that human evolutionary changes happened contemporaneously in different regions of the world, producing different races, albeit with the careful caveat that some genetic exchanges occurred between regions, thus still allowing for some similarities among the human species (Cela-Conde and Ayala 2007: 253). This account lends credibility to all kinds of theories about cultural relativism, racism, and similar theories.

According to this theory, *Neanderthal* supposedly evolved into *Cro-Magnon* in Europe, while *Homo erectus* evolved into modern Asians, and *Homo ergaster* into modern Africans. This assumption underpinned the definition of race along geographical lines: Caucasians (Western Eurasia), Black Africans (Sub-Saharan African), Mongoloids (east Asia), South Asian Aborigines (Southern India), Amerinds (Americans), Oceanians, and Australian Aborigines (Wells 2002: 17).

In all these theories, the environment plays an insignificant role and therefore these explanations seems to run counter to standard Darwinian thought, namely that biological evolution is based on adaptation and selection. This may be because it has been rather difficult to pinpoint a specific new practice or technological achievement in the repertoire of early *Homo sapiens* that could explain speciation. All the archeological records show is evidence of the relatively sudden emergence of symbolic capability. But to get to this point, early *sapiens*, like all other animals, would have had to invent a new practice that could subsequently be refined through natural selection, and which would have caused them to become *Homo sapiens*.

Recent studies in genetics also make all the above evolutionary scenarios highly unlikely. As it turns out, human genetic variation is way too small to support such explanations. Notably, genetic studies of different humans groups show that the majority of genetic differences—85 percent—are found *within* populations, with only 7 percent variation differentiating populations within a single race, and only 8 percent variation differentiating the so-called contemporary "races" mentioned above (Wells 2002: 17). So, for instance, if all humans apart from the population of Copenhagen were wiped out, 85 percent of the total human genetic diversity would be retained.

Genetic studies also show that the world has not been such a genetic melting pot as we may once have thought. People have, on the whole, conservatively stayed in the places where their ancestors lived. But this does not mean that they produced culture separately in their respective locations, or that the Europeans were the first to develop artistic expression and complex culture.

We know this because genetic studies have confirmed the Out-of-Africa hypothesis. As Gamble writes,

> The clinching argument came in 1987 with the publication of a genetic tree (Cann et al. 1987) based on the inheritance of mitochondrial DNA (mtDNA) that is restricted to the maternal line. Their sample of 148 women from the world's five major geographical populations demonstrated that Africans had the greater diversity, and hence were older, than any other biological group (Gamble 2007: Kindle Loc. 921–24).

So, the human genetic tree leads straight back to Africa within the past 100,000 years. From here, *Homo sapiens* migrated round the Indian Ocean coastline and on to Australia, which they arrived around 70,000 years ago. Finally, around 50,000 years ago, the early humans reached Europe. We know from the large amount of rock-art left by these early humans in Australia that they were already painting by the time they came out of Africa (Oppenheimer 2003) and hence that they already had a symbolic cultural system. As Burroughs writes "The extraordinary range of rock art in Australia adds great weight to the idea that artistic creativity was part and parcel of the intellectual capacity of modern humans that migrated out of Africa" (2005: Kindle Loc. 1652–53).

We also know that the human genetic heritage is derived from a population that may have been as small as 2,000 individuals and probably not more than 10,000 individuals (Oppenheimer 2003: 46). And, since early humans arrived in Australia with a complex artistic culture (Balme et al. 2009), we must assume that these 2,000 to 10,000 individuals were the ones that "invented" culture.

So the accumulated knowledge of archeology and recent discoveries in genetics about human origins does not fit any of the presented scenarios of human evolution. Instead, human evolution shows long periods of stasis interrupted by periods of relatively swift and sudden change, and, in all these cases, technological improvement occurred *after* these transitions (e.g. Oppenheimer 2003). It seems that evolution toward intelligence and culture is not a general, multiregional process, and therefore that there must have been some adaptive mechanism that made it possible for a small group to survive in Africa. *Homo ergaster* must somehow have experienced speciation into *Homo sapiens*.

We therefore have to search for some kind of universal feature in the panoply of human cultural diversity—the feature that must have triggered the early transition into *Homo sapiens*. And indeed, there is one more universal, apart from fire, cooking, and grooming, that has so far been overlooked: namely kinship. All human groups have kinship, and they all have rules about kinship—who to marry and who to avoid. No other animals have such intricate rules about whom one can mate with.

Kinship is arguably the most basic form of human interaction. Not surprisingly, it was a huge concern among early anthropologists: every self-respecting monograph had a section about kinship. Anthropological knowledge is, therefore, essential to understanding the evolutionary circumstances that led early humans to invent kinship. Yet anthropology as a field has not shown much interest in the subject of human evolution. As Chapais argues,

> one would have thought that anthropology would be eager to uncover the principles that enable human beings to apply symbolic thought to our environment, especially given that we now know that all human societies originated from the same basic building blocks and are thus constructed upon the same basic structure (2008: 3).

Since kinship is a practice carried out among people it did not leave any concrete traces in the archeological record, but all the paraphernalia normally associated with kinship in anthropology is present, though these are otherwise only used by archeologists as evidence of the abrupt appearance of a fully developed human symbolic culture. Kinship is the clue to the last transition into *Homo sapiens*.

Consequently, we have to search for a new way to understand the dynamic that produces cultural differences, rather than just accepting that humans prefer to "naturally" congregate in culturally similar groups. Cultural differences must, therefore, have been part of the human evolutionary heritage, a product of human interaction, and not just a state of affairs characterizing different groups. Otherwise it is difficult to see how the original small group of migrating humans could have produced today's enormous cultural variations.

Cultural differences must have emerged from some form of basic symbolic capability acquired by early *Homo sapiens* and which accompanied the different human groups into their new territories. Looking at human evolution from this angle means that we have to give up the truism that humans gradually evolved by using tools or learning language to communicate, thereby eventually stepping into a fully symbolically constituted universe (e.g., Gibson and Ingold 1993). The human evolutionary sequence strongly speaks against this scenario, since it shows long periods of stasis interrupted by relatively swift changes. It was during one of these changes that culture appeared.

Prehuman species had to adapt to dramatically fluctuating weather caused by recurrent ice ages, which each time posed new challenges to their populations. Therefore, this book uses an ethnographic approach to investigate how the different human species adopted new strategies to obtain food and exploit the changing environment, and how in the process they developed the characteristics of the present human population: bipedalism, cooking, and the use of symbols.

EVOLUTION AND THE FUNDAMENTALS OF THE SOCIAL SCIENCES

If we accept that all social groups share a common evolutionary background that contained the seeds of cultural variation, this has great implications for the social sciences. Unless our theoretical understanding of human society in all its shapes and forms can be linked to human evolution we are just fumbling in the dark, guessing about what drives the nature and direction of human society and destiny. Our common heritage must be incorporated into theories of social and cultural organization and change; otherwise, these will remain based on speculation about fundamental human nature and societal mechanisms. In sort: our whole understanding of the human condition and what shapes our lives revolves around our understanding of what caused human evolution.

This may seem a bold statement to some. After all, the world has changed profoundly since early humans emerged somewhere in Africa. Many would claim that human lives are now so diverse and complex that they have little to do with early evolution. New conditions need new theories. But such views rest on the assumption that human beings are completely unconstrained: once they had appeared, they were magically able to transcend all barriers, natural or divine. To many, evolution was what liberated humans from their dependence on the environment, in contrast to all other animals. Something set humans apart—but what? Some haphazard genetic mutation? Magic? Nobody knows for sure, but this has not stopped many from speculating freely about the origins of the human condition. Anything goes, it seems, as long as it is vaguely plausible.

At this point, some might argue that it would not be possible to manage contemporary complex societies without the help of economic theories. And obviously these theories are not the least concerned with human evolution or cultural diversity. But most economic theories assume that pan-human conditions exist despite cultural diversity, thereby claiming a kind of a default, generic human condition upon which human differences are built. No explanations are offered: we are all just "*homo economicus*." Many economists' belief that economic behavior is universally constituted is essentially an—albeit unintentional—evolutionary claim.

Ever since Adam Smith published "*The Wealth of Nations*" in 1776, "the economy" has been seen as a vital issue in all contemporary societies. For Smith, the creation of wealth was a central issue in the running of contemporary society, and this in turn required specialization and therefore trade. Contemporary society was reduced to a marketplace, and this was seen as the central process through which a society allocated its wealth and resources. For Smith, "the market" secured the best allocation of resources and their optimal use so that the total wealth of society was maximized to the fullest. It was a market based on competition and self-interest and together these mechanisms would provide what people needed.

Self-interest has subsequently been enshrined as a fundamental assumption in economic theories. You would think that such an essential claim about human sentiment and its influence on social organization would lead to some biological considerations. But most economics traveled another path (Beinhocker 2006: 17). They based their understanding of the market on physics, and especially on the physics of motion and energy.

A general assumption was that when people attempt to maximize their happiness in a world of differing utilities and finite resources, they will inevitably trade their way to a point of market equilibrium. As a result, the economy has been portrayed as an equilibrium system that moves from one such point to the next, propelled by shocks from external events like the invention of new technologies or changes in consumer preferences (Beinhocker 2006: 17).

As in physics, such equilibrium points could be mathematically calculated. All one had to do was to assume that people are logical and consistent in their behavior. In these theories consumers are believed to be rational in their economic choices. From these assumptions twentieth-century economists created rigorous mathematical models for describing the workings of the economy. Policymakers and bankers all rely on these models in their daily work.

It has even been claimed that these economic assumptions have a pan-human significance and can be used to describe economic systems historically and globally. In this way these economic theories can be seen as based on the assumed existence of some sort of basic human condition, although this condition is never explicitly stated.

Those who advanced such a view of the universal nature of self-interest became known as "formalists." They argued that the formal rules of neoclassical economic theory, derived from the study of capitalist market societies could be used to explain the dynamics of pre-modern economies as well. For them, all economies involve the rational pursuit of, access to, and use of, scarce resources by self-interested, maximizing social actors, and therefore formal economic rules could be used to explain them. Harold Schneider became the dominant figure in this school. He posited the universal nature of self-interest as follows: "The unifying element among . . . formalists is, . . ., the partial or total acceptance of the cross-cultural applicability of formal [microeconomic] theory" (1974: 9).

Herskovits made the pan-human applicability of these assumptions even clearer. In *The Economic Life of Primitive Peoples* (1940) he argued that scarcity is universal and so is maximizing behavior on the part of the individual. For him, it was only the cultural matrix within which these occurred that varied, but self-interest was the means that were everywhere applied to achieve different ends.

Perhaps these economic theories could defy the ultimate test of evolutionary relevance, despite their claims to universal self-interest, by putting forward a coherent explanation of contemporary behavior, or proving that evolutionary forces are not significant in contemporary human lives. In the last instance they would have to base their claim on a "before" and an "after" scenario—for example, investigating when evolution stopped having an influence on economic behavior, allowing self-interest to prevail, and explaining what caused this transition. Yet mostly, self-interest is just taken for granted as a basic pillar of the human condition.

However, at the beginning of the 1900s, a dialogue arose among economists (Léon Walras), physicists (Joseph Bertrand), and mathematicians (Henri Poincaré), for example, between the economists and those who supplied their mathematical models. While the mathematicians found that it was praiseworthy for economists to become mathematically rigorous, they also thought that the assumptions the economists used in their equations meant that they were throwing reality out of their models. They were particularly concerned about the assumption of the unlimited foresight of economic actors (Beinhocker 2006: 49). In their view, the economists assumed that humans are like superhuman robots that live in theoretical worlds that are much simpler than the world real people live in (Beinhocker 2006: 51).

According to the mathematicians, the economists seemed to ignore the unrealistic assumptions about human behavior that their models were based on. They seemingly found it unnecessary to take the real world into account when making predictions about it. Still, the economists stuck to their guns, arguing that as long as the theories made correct predictions nothing else really mattered (Beinhocker 2006: 49). You could just as well claim that as long as day follows night, what do we care if giants are dragging the sun across the sky? Still, most people would probably like to know where the giants came from. This version of economics is basically fantasy construed as reality, and then presented as social science.

These economic theories are part of a larger theoretical body called cultural materialism. Here, the claim is that the constitution of culture and society are dependent on their material basis, mode of production, or technological level. In this line of thought, cultural differences are just obliterated, or simply seen as a reflection of material forces that propel human societies forward to ever more complex stages. Accordingly, we can have all kinds of, for example, artistic expressions, different uses of symbolic systems, and different beliefs, but what really matters has always been the technological base.

This, of course, makes all mice gray, and leaves little room for understanding cultural variations, which are merely seen as different *stages* of a common evolutionary process. Each stage has usually been characterized by its level of technological accomplishment, which is probably due to the prominent role of the "human the tool-maker" idea in the study of human evolution. In this tradition, technological knowhow or productive skills are the defining features of each society. This idea is found in the way cultural evolution has been characterized as progressing from Stone-Age to Bronze-Age to Iron-Age; from agrarian societies to industrial societies to information societies; or from primitive accumulation to feudal accumulation to capitalistic accumulation, ending up with communistic accumulation, which completes the circle—equality at both ends.

These economic models may seem to work short-term, and this is probably due to human behavior being pretty consistent over shorter periods of time. But from the perspective of the "longue dureé," we see that major economic centers have historically shifted from ancient Greece, to Rome, to London to New York, seemingly evading the logic of these models. It is after all, difficult to see how self-interest and rational choices caused such major upheavals, that to all intents and purposes have not benefitted those in the abandoned centers, who (according to economic models) caused the shifts in the first place. Historically, there is, therefore, not much that vouches for these models, although in the short-term they may seem able to make crude approximations.

Yet economists still think that their assumptions are valid as long as their models produce reasonably reliable results. They are unconcerned about whether their assumptions hold water seen from an evolutionary perspective because they find no evolutionary claims in Smith's conception of "the market" or in later economic models. At the same time, these economic theories claim universal applicability across time and space. Hence they also make claims about the basic human condition, and therefore also about the evolutionary nature of human capabilities. It is economists who have equipped humans with self-interest, which is the default mode, purported to drive markets.

But such a drive must come from somewhere. We need answers to the questions of where, when, and how it emerged—and what happened, or did not happen, elsewhere. If self-interest is assumed to be a basic human condition then it must also have played an evolutionary role in the creation of *Homo sapiens*. Otherwise, the notion of human self-interest as an innate drive is just free speculation. In evolutionary terms, at least, the economists are completely of the mark.

I am not the first to challenge the pan-human applicability of major economic models. A group of scholars called "substantivists" opposed the formalist school and its application of the market model to all societies. The substantivists are part of a larger school of social science thought called social relativism. Social relativists claim that the "economy" is submerged in cultural contexts and that it was only in the West that "the economy" emerged as a separate social sphere. For the substantivists, formalist economic thought was only applicable to capitalist societies.

Karl Polanyi was the dominant figure among the substantivists. In *The Great Transformation* (1944), he analyzed the emergence of a new type of economy, market capitalism, that he posited first appeared in England during the early nineteenth century and then spread to the rest of the industrializing world. According to Polanyi, this new economy was unique in that it was disembedded from the social matrix. In previous times, "man's economy. . .[was] submerged in his social relationships" (Polanyi 1944: 46). People gained access to land and labor through ties of kinship (birth, adoption, marriage) and community.

For Polanyi the difference between the industrial capitalist economy of the West and both contemporary and historic premarket economies was one of substance, hence the name "substantivists." The latter further argued that different types of economies did not lend themselves to analysis using the same methods. For the substantivists, the "economy" is organized according to cultural variation, and there is no pan-human mode for how this economic organization takes place. Right up until the time when a group reaches the market economy stage, peoples' lives are thought to be organized by culture. In this line of thought, cultural relativism is just taken for granted as a human specialty in which differences pass unquestioned.

In general, cultural relativism presents culture as integrated and closed units, separate from one another. As Pagel observes, "Why do we humans have a tendency to form into small tribal groups rather than living as one large and homogeneous society?" (2012, vii). But instead of merely observing the existence of cultural variation (and maybe cultural relativism), we should also try to explain it.

According to the substantivists, cultures operate solely according to internal laws that are specific to each one. This assumption effectively obviates investigation into the evolutionary origins of culture. Indeed, it has become an anthropological truism that culture shapes our lives and makes us different from each other. Once "out of nature," humans never looked back. Culture came to be seen as a system unto itself. Its sheer complexity and diversity seem to preclude any search for its evolutionary origins. It seemed that all that could be done was to study culture in its different regional and local manifestations; for example, try to understand human diversity and the many different ways people organize their lives. Accordingly, modern social sciences can start from here.

This view has been around since Boas, whose weighty legacy in anthropology emphasized the uniqueness and originality of every culture, and whose school of thought was therefore exclusively concerned with cultural *differences*, rather than the similarities between societies (Chapais 2008: 135). Boas and his followers rejected the notion that anthropology should be concerned with the formulation of general laws governing culture and society. Cultures had to be understood from the inside out, and this naturally precluded any research into the evolutionary origins of CULTURE, writ large, as the result of adaptation in the classical Darwinian sense.

This view was also endorsed and promoted by Geertz, who wrote that man is an animal suspended in webs of significance he himself has spun. For Geertz, humans are bound up in a series of symbolic representations that serve to generate the meanings that guide individuals through life. Based on these representations, people generate ideas and learn to act with other people. There can, of course, be distortions in such representations, which will lead to reconfiguration of the representations, and so on (Geertz 1973).

Taking this one step further, recent post-modern theories would have it that humans can only understand and relate to their environment through their linguistic categories. According to this view, meaning only exists within a system of signifiers, and through each sign's propositional value in relation to other signs (e.g., Hall and Open University, 1997). Within this view, even nature is constructed by language.

According to cultural relativism, humans are at all times caught in their symbolic representations. But, as Wheeler points out, these accounts of culture and language omit how humans came to use language and other symbolic instruments in the first place. If environments and organisms are not real, but are only constructed in language, then evolution simply could not occur (2006: 23). For evolution to occur, some kind of reality must exist. In a fundamental way, humans must be designed to interact with that reality. In evolutionary terms, then, cultural relativism is also off the mark. Otherwise one would have to believe that culture was magically conjured up out of thin air.

To sum up, according to the formalist/materialist approach, the human economy is driven by greed, whereas for the substantivist/relativist school human lives and their drives are organized by culture. This is an almost insurmountable dichotomy in the social sciences, and it has been debated for centuries: is society the result of lone individuals interacting, or are individuals and their interactions determined by their cultural rules? Whichever view any given researcher espouses usually seems more a question of personal conviction than explanatory power. Both theories claim supremacy. You have to choose between cultural relativism and cultural materialism with little to guide that choice except an endless and circular debate.

However, at the root of the substantive/formalist and relativist/materialist debate we find some gnawing problems. For relativists, culture is just "there" and we do not need to bother with how it got there. For materialists, trade and exchange produce culture. But, in fact, exchange must itself be a derivative of culture since some form of symbolic thought is needed in order to compare products and establish value. Culture must therefore precede exchange, otherwise we must assume that things and objects carry a "natural" symbolic value all by themselves which can be realized in exchange, thereby purportedly producing culture. This would indeed be a brave new world with completely new rules of interaction, not just for humans but for all other animals as well. Neither of these approaches makes sense from an evolutionary point of view, as I shall show in this book. Both views are based on undocumented ideological assumptions freed from evolutionary concerns.

Purged of dubious assumptions, the above mentioned theoretical debate leaves us with two rather broad statements about the human condition: people have a propensity to trade, and they congregate in different forms of cultural organization. This leaves us with two basic conditions for human life: trade and cultural differences. These must, therefore, be the defining characteristics we need to search for in human evolution. Both must have played a significant role in our transition into modern humans. This means that we have to search for a situation that forced early humans to begin trading—a circumstance that made trade a matter of survival. We need to identify an adaptive change made by early hominine populations that linked these two different characteristics of present human populations.

On the face of it, there doesn't seem to be any pan-human way of doing the same thing. This is the backbone of cultural relativist thought. But this is the micro-view of cultural diversity. What should concern us from an evolutionary macro-perspective is that all human groups engage in culturally diverse forms of organization. But the commonality of this form of organization has been obliterated from view by cultural relativists in favor of an exclusive focus on the cultural units themselves.

If you think about it, the human population has chosen a strange way to organize itself. In contrast to all other animals, human populations do not spread according to the availability of resources. And all humans congregate in distinct groups where individuals are basically doing the same things, just in different ways. Every human individual belongs to a group with its own symbolic make-up and organizational structure. This form of cultural organization is universal—not a single group of people escapes it. Groups can share the same technology and still be very different in their organizational and symbolic constitutions, as we see in present day Papua New Guinea, for instance. Even human groups living in proximity are everywhere characterized by heterogeneity and division. Whatever else we are, we also belong to symbolically constituted communities. This is not a new or recent phenomenon: it has been our human condition ever since we turned into *Homo sapiens*.

So far, the human population has nowhere transcended this predicament, not even by intelligence, rationality, and creativity, which are otherwise claimed to be the prime movers in human history. There is something more than mere rationality, creativity, and intelligence to this way of organizing human lives: namely, cultural diversity must have been evolutionarily advantageous since it has always been with us—for better or worse. In principle, all human groups might just as well have had the same culture. But they do not, and this is the truly remarkable feature of the human population. Despite that, it is usually left unexplained.

Human evolution and cultural diversity must therefore be linked, and it is crucial to investigate the relationship between these two processes if we are to understand the human condition. It is the default mechanism guiding all subsequent human cultural expressions and organizational forms. It is this relationship we need to understand before we can say anything meaningful about social issues and processes, and all social theories must be tested against this fundamental default mechanism in human lives.

I therefore posit in this book that these two spheres of the human condition are interlinked, or presuppose, one another. Without cultural differences, there would be very little incentive to trade. And according to Paul Seabright (2004), cooperative trading between nonrelatives is a uniquely human activity. Trade, in turn, enables specialization—it thrives on differences. Trade, therefore, depends on difference, and the two are intrinsically linked.

This view elevates the creation and evolutionary roots of cultural differences to our major theoretical concern. Culture is then the human evolutionary watershed. This is a view held by many theorists who find that culture is the turning point in the history of human evolution and also the most basic characteristic of the human condition (see, e.g., Carrithers 1990; Klein and Edgar 2002). Some think that at a certain point in time, humans stepped outside nature and took a huge stride away from biological adaptation and toward cultural innovation (see Wells 2010: 112). As Lewin and Foley put it: "human evolution has sometimes been thought of in terms of a transition from biological Darwinian mechanisms, to cultural evolution, with a species freed from the constraints of natural selection" (Lewin and Foley 2004: 458). This shift has been called the "great leap forward," with some proponents even claiming that "[c]ulture. . . freed the human race from the pressures of natural selection: We made clothes rather than growing fur and built better weapons rather than becoming stronger" (Cochran and Harpending 2009: 2).

However, these ideas about human evolution simply assume that humans just happened to become different from all other animals. Culture made humans different and released them from nature's iron grip. Ever since, culture has become a human given and it is commonly seen as the human condition par excellence: "[c]ulture is the trump card that enables humans to adapt" (Wrangham 2009: 11).

But actually, human culture, in whatever form it presents itself, is a kind of evolutionary contradiction. Not only do humans have cultures, these are also different from one other. In all other animal species, variations are always connected to differences in ecological niches; only in the human population do we find biologically similar groups that are behaviorally different from one another. So, culture is a result of a common evolutionary process, but at the same time this process also made people culturally and thus behaviorally different. And it is perhaps this characteristic that separates the human race from all other species.

Difference and variation are so fundamental to human life that they seem to be the most important characteristics of the human population. But it seems we have become so accustomed to cultural differences that we take them for granted instead of asking what purpose they serve, what kind of dynamic created them, and what mechanisms keep them running. In evolutionary terms, differences and variations must obviously have been essential to survival. To be able to communicate with symbols is one thing—and it is difficult enough to account for this from an evolutionary perspective—but that this same ability should somehow also produce differences is another thing entirely.

It seems that both culture and cultural differences should be seen as part of the same evolutionary equation, so that within any culture there is also a mechanism for producing differences. Seen thus, culture and cultural differences must be part of the general process that brought culture into being in the first place. From an evolutionary perspective, humans are the outcome of general mechanisms that apply equally to all other organisms, hence culture and variation must be seen as adaptive mechanisms. Otherwise the human population would have similar cultures the world over.

But there is a major problem here. Having established that human are characterized by culture, we must instantly recognize that culture is the big unknown, and indeed social scientists define it pretty much any way they like. To make matters worse, those who study evolution are generally not experts on culture; and those who know about culture, notably anthropologists, do not normally care much about evolution, and in any case they seldom agree about what culture is or does.

Due to the uncertain nature of culture, many in anthropology now find that the notion of culture has become too broad to explain anything. As Trouillot observes, "[c]ulture now explains everything: from political instability in Haiti to ethnic war in the Balkans, from labor difficulties on the shop floors of Mexican *maquiladoras* to racial tensions in British schools." (Trouillot 2002: 37). Or, as Tucker complains, culture "denotes both everything and nothing. Culture appears stripped of all content and yet is seen as determining everything" (1997: 1). Such critiques give us the chance to start afresh, and to incorporate the conundrum of evolution into our understanding of culture.

So instead of trying to define culture by investigating our contemporary cultural organizations and their multitude of cultural expressions, we should try to find *the elementary form of culture* that had an evolutionary potential and that can explain differentiation and trade between groups. Since it is such a defining human phenomenon, cultural variation must have offered an evolutionary advantage in this venture, just as it must have played an adaptive role for early humans.

This book argues that cultural differences must have emerged from the same basic mold based on the same basic symbolic capabilities acquired by early *Homo sapiens* and which accompanied the different human groups into their new territories. For this to come into existence, humans must have shared an elementary structure of meaning. It is this structure that enabled early *Homo sapiens* to extend their ecological niche, eventually colonizing the entire globe and forcing former species of the homo lineage, *Neanderthal*, *Homo erectus*, and *Homo ergaster*, into extinction.

So here we are again, on the threshold to *Homo sapiens* and the transition into a culturally constituted universe, or perhaps we should just call it symbolically constituted? In both cases, we still need a mechanism that made this possible, a change in behavior that carried the seed of symbolic thought, a new way of doing things that had symbolic potential. And, as I have already hinted, this practice is kinship.

How could something so obvious have gone unnoticed for so long? I believe there are several reasons for this. Since the dawn of archeology, evidence of human evolution was found in a rather haphazard manner and in very different locations. And for a long time it seemed that humans had evolved simultaneously in different places, and that each of these places provided a kind of evolutionary record with "layers" of different hominids occupying the same places at different times. Let me start by setting this record straight.

It would be easy to interpret these findings using existing theories, most of which posit that evolution is a cumulative process. This has led to a search for a kind of turning point from which *Homo sapiens* would eventually evolve. But we now know that evolution takes place in bursts followed by long periods of stasis. So we have to identify these periods in order to establish the role played by kinship.

In practice, this means that we have to compare these periods of evolution and stasis against the ethnographic record. First of all, we must investigate studies about kinship, and then we must look at studies that link kinship to exchange. We can then relate this to systems of exchange, as we find them in Papua New Guinea, for instance; and then to studies that link exchange to ethnicity. This will give us a more holistic picture of human kinship and evolution, and an explanation as to why the human condition is characterized by cultural differences and exchange/trade.

Subsequently, these findings will be used to qualify the most basic assumption of the social sciences. To accomplish this, we first have to scrutinize existing assumptions about human evolution. From an evolutionary point of view we need to sort out the many popular ideas about human evolution and its sequence.

Chapter One

The Record of Human Evolution

Kinship is mostly seen as a social institution that has taken widely different forms in different societies. That may explain why the evolutionary and adaptive role played by kinship has so far been overlooked.

Instead, the traditional sequence operates with four key events in human evolution: (1) coming down from the trees; (2) learning to walk upright; (3) brain expansion; and (4) the invention of culture. Often, this sequence is depicted as a kind of heroic tale, as in the following description: "The climate changes, the forest shrink, and the hero is cast out on the savannah where he faces new and terrible dangers. He struggles to overcome them, by developing intelligence, learning to use tools, and so on, and eventually emerges triumphant, recognizable you and me" (Lewin and Foley 2004: 20).

As Elliot Smith, an early twentieth-century English anthropologist put it, it is: "The wonderful story of Man's journeying towards his ultimate goal" (in Lewin and Foley 2004: 20). Such stories picture human evolution as steadily and inevitably progressive—a program of constant improvement in which each stage lays the foundation for the next. Seen from this point of view, *Neanderthal* could be understood as a proto-type of later *Cro-Magnon* man, who would, in turn, have needed little incentive to unfold his evolutionary potential.

Such a view of evolution makes human evolution seem uncomplicated and inevitable. But it also takes for granted much that still remains to be explained. Human life is, after all, a very strange phenomenon in the animal kingdom and it seems only reasonable to ask how we became quite that strange. Answering this question is not an easy matter. Historically, beliefs, ideas, and theories about human evolution have abounded. Every society has its very own creation myth. Most great religions are founded upon the idea that humans resulted from some form of divine creation. Scientists have also grappled with this question since before Darwin's times. Often these different views of human evolution become mixed up, both in public debates and in the scientific community. Even hard data can be deceptive. Myths, ideologies, and truth are often hard to separate. Preconceived ideas about evolution, culture, and the human condition have led people to search in all directions in an attempt to explain these phenomena.

Historically, several major misunderstandings and assumptions have confounded our understanding of human evolution and notably the appearance of culture. Some have even gone so far as to ban the study of human evolution or deem it useless. In anthropology, for instance, English anthropologist Evans-Pritchard expressed distaste for the endeavor to search for the origins of a given human trait and try to trace its subsequent evolutionary trajectory. He wrote: "To my mind it is extraordinary that anyone could have thought it worthwhile to speculate about what might have been the origin of some custom or beliefs, when there is absolutely no means of discovering, in the absence of historical evidence, what was its origin. . . for which no evidence was, or could be, adduced" (1965:104–8).

Certain linguists were, similarly, of the opinion that it was a complete waste of time—indeed almost disreputable—to try to explain the particularities of human language from an evolutionary point of view. In 1866, the *Société de Linguistique* of Paris declared a moratorium on the topic (Kenneally 2007: 7), declaring that such an endeavor was completely futile and banning the search for the origin of language.

But their lack of concern about the dynamics that brought their very scientific topic into existence also made it practically impossible for them to define what could plausibly be said within the confines of their respective disciplines. And although at that time, such a stance may have seemed appropriate, nowadays the archeological evidence is piling up and genetics has supplied the material to show the very recent common origins of the human species and its subsequent migration route out of Africa. It is therefore essential to understand the evolution of culture in order to qualify any statement at all about the human condition.

THE ARCHEOLOGICAL RECORD

Early finds in archeology did not initially further our understanding of the evolution of culture. These finds occurred rather haphazardly and at first they did not prove much about human capabilities other than showing clear signs of human ancestors who had undergone some kind of evolution. But then fossils started to appear all over the world. These finds seemed to prove the general potential for an intelligent creature to emerge as the result of some general evolutionary tendencies. They also appeared to shed light on contemporary human diversity.

Debates began in the nineteenth century, when archeological finds in Europe showed that other hominid species had existed in the past. In 1856, a skull was discovered in the Neander Valley of western Germany. Initially the skull was thought to be the remains of a malformed modern human. But a year later it was presented to the world of science at a meeting of the Lower Rhine Medical and Natural History Society held in Bonn. In 1864 it was named as a species and called *Homo neanderthalensis* by William King. Later, remains from other individuals of the same species were uncovered. In 1887, two complete skeletons were found in a cave near Spy in Belgium, and more were found in France in 1887, 1908, and 1911. It became increasingly clear that the hominid lineage had evolved over time (Wells 2002: 34).

But what really corroborated the idea of a general human potential for evolution were discoveries of early human species in other parts of the world. In 1890, Eugène Dubois working for the Dutch East India Company found an odd skullcap in Java. He called it *Pithecanthropus erectus*, the Latin name for "erect ape-man." In the scientific community it became known as the Java Man. Then other similar fossil remains turned up in China, which became known as Peking Man. Together these finds showed that other, now extinct, species of hominids had existed. Then, in 1921, a Neanderthal-like skull was found in Northern Rhodesia, proving that Africa also had its own early human species. These diverse forms of early human were all believed to possess the potential to evolve into modern *Homo sapiens* (Wells 2002: 35).

As other finds indicated, *Neanderthal* and his contemporaries were fairly sophisticated: they were hunter-gatherers, they used tools, and they could control fire. *Neanderthal* consistently used a particular type of tool that consisted of a biface stone that was more or less flaked on both surfaces (Klein and Edgar 2002: 102). This is called the Acheulean tool complex, named after the first place where these tools were excavated, for example, in the gravel pits near St. Acheul in France (Klein and Edgar 2002: 105). This tool has also been called a "hand-axe" because of its drop-like shape, with no handle, implicating that *Neanderthal* used it as a skin and bone axe to butcher their larger game animals.

Its use has been widely debated, with suggestions as to its function and shape ranging from a waste product of flake-making, to an early prototype "Swiss army knife" that could be used for a variety of jobs, and even a "killer Frisbee" that could be hurled at grazing animals, making them tumble over for an easy kill. There thus appeared to be an intimate relationship between this species and its technological knowhow, suggesting that tool-making and human evolution were interlinked, and that, accordingly, the evolution of human culture was linked to the mental challenges that tool-making was assumed to pose.

However, it is also clear from the fossil record that these forms of early human species became extinct at around the same time some 30,000 years ago, or rather had seemingly evolved into a fully modern human species in what Klein and Edgar (2002: 204) call a geological blink of the eye. After 30,000 years ago we find only the remains of fully modern humans named after the rock shelter in France where some of the earliest bones were found in 1868. They became known as *Cro-Magnon* Man. This new species was different: its members were taller, they had longer limbs, they made more sophisticated tools, and they painted.

There had clearly been a radical change in *Cro-Magnon*'s technological capabilities. His blades were long, elegant, and sharp-edged. He used drills, chisels, and different kinds of points and scrapers. He had invented the spear-thrower, and made nets to trap small game. *Cro-Magnon* also made many forms of jewelry such as necklaces, pendants, rings, bracelets and anklets, and he used different materials to produce these, such as ivory, shell, bone, stone, and animal teeth. Many of these implements displayed carved decorations. Most of all, *Cro-Magnon* painted impressive cave paintings using red and yellow ochre (Stone, et al. 2007:,43–44).

Many of the raw materials they used for making their tools and jewelry were obtained from great distances away, suggesting that they were involved in some kind of long distance trade (Stone, et al. 2007: 43). They buried their dead and they gave them gifts that were buried with them. If nothing else, this relatively swift transition corroborated the idea that it was *Cro-Magnon* who took the huge step into a symbolically constituted universe—seemingly in the blink of an eye.

Thus, the early archeological records seemed to show that our human ancestors had evolved separately everywhere, maybe as part of a general evolutionary thrust that produced similar species in different places. Humans seemingly had an inner potential that was destined to emerge at some point in time, leading them into a fully constituted symbolic universe that was at once similar and different in its various manifestations. From then on, archeology could continue to search for "the missing link," and anthropology could stick to trying to understand human cultural diversity.

However, this scenario leaves us with a conundrum. There were similar technologies covering a vast geographical area, extending from Europe to China and on to Java that were supposed to propel human evolution locally. This suggests a general evolutionary trend whereby human populations go through similar technological stages everywhere. At the same time, this universal connection between technological skills and human evolution is also believed to be the dynamic responsible for producing different cultural formations in different regions. But it seems difficult to reconcile this common technological complex with a sudden transition into different symbolically constituted cultural orders.

There is another minor inconvenience in this scenario of parallel similar, yet different, human evolution. A predecessor, or a "missing link," to *Neanderthal* in Europe or to his equivalents in Asia was never found. In these regions, human evolution seems to have started with *Neanderthal*, Java Man and Peking Man, all of whom gradually turned into *Homo sapiens*. This is by any standard a very shallow account of evolution. Neither was a species ever found that remotely resembled *Neanderthal* in Europe or a predecessor to Java/Peking Man in Asia, nor any species, human- or ape-like, that could possibly have evolved into *Neanderthal*. Europe and Asia would appear to be archeologically devoid of such creatures. Was there perhaps a species working undercover in Euro-Asia?

When the "missing link" was finally found it was in Africa, by Professor Dart from the University of Witwatersrand in South Africa. In 1924, he was investigating the *Taung* cave near Johannesburg, and there he unearthed a child skull with canine teeth much smaller than those of an ape. More importantly, the skull also had a *foramen magnum* (the connection between the spinal column and the brain) that was inclined downward as in modern humans, instead of backward as in apes. This creature had clearly walked upright. In 1925, he concluded that the skull was the remains of a new species and he called it *Australopithecus africanus*, the Africa Southern ape, better known as the *Taung* child (Wells 2002: 36). Dart had found the first clear evidence of a missing link—in Africa, a very long way away from the creature's descendants in Europe and Asia.

If nothing else, this find signaled a flurry of human-origin research in Africa. It also raised huge questions about the multiregional origins of humanity. The studies that followed focused their attention on the Africa Rift Valley system. This valley is formed by the movements of the tectonic plates, and it stretches from Eritrea in the north to Mozambique in the south. It is dotted with big lakes, volcanoes, and mountains and it influences the weather in the region. It has been a place of intense ecological activity for the last 20 million years (e.g., Wells 2002: 37).

It was in this region that the Leakeys conducted their world famous excavations. In 1959, they were working at Oldovai in northern Tanzania, and there they stumbled upon a skull. After three weeks in the laboratory they could announce the discovery of the first *Australopithecus* in east Africa, but it was the skull's age that was the real news: it was more than 1.75 million years (Ma) old. This nearly doubled the presumed age of the human lineage at that time. Subsequently, many more examples of this species were found there, suggesting that Africa was the locus of our original human ancestors (Wells 2002: 38).

Later, in the 1960s and 1970s, fossils of *Homo erectus* began to appear in east Africa, which in its African variant became known as *Homo ergaster*, with the oldest example from around 1.8 Ma. There was no longer any doubt that humans originated in Africa, and that they spread out from there to become *Neanderthal*, Peking man and Java man. Eventually this lineage could be traced back for more than 5 Ma to the recently discovered chimpanzee-like *Ardipithecus*, giving humans a long line of ancestry (Committee on the Earth System Context for Hominin and National Research 2010: 18).

But this story tells us nothing more than that humans originated in Africa and that they spread out from there. It does not tell us anything about the transition into modern, cultural *Homo sapiens*. There is, after all, a long stretch of time between the early hominids found in Africa and the *Neanderthal* skeleton found in France that dates back 40,000 years. In principle, practically anything could have happened in the meantime.

These finds did not preclude that humans had made a sudden and abrupt transition into a symbolic universe in several different places at roughly the same time. And they certainly did not preclude that *Neanderthal* somehow transformed into *Cro-Magnon* not so long ago. Up until then, the archeological record corroborated such a hypothesis. In this way, early archeology underpinned the idea of a continuous and general evolutionary process that would produce the same results everywhere, given enough time.

Home erectus in all its variants seemingly turned into *Homo sapiens* all over the globe. Because the transition to *sapiens* occurred in different geographical locations, this story also seems to account for why *Homo sapiens* behaved and organized differently in different places. So, the evolution of culture was seen as the inevitable end product of some kind of cumulative process in which humans gradually developed their mental capabilities, either through producing tools or through social interaction. Either way, they ultimately became highly intelligent beings.

This idea of a general evolutionary process was once so taken for granted that it became standard textbook material. It was even included in the classic anthropological textbook by Roger Keesing *Cultural Anthropology—A Contemporary Perspective* (1981), which presents the standard story about human evolution. According to Keesing, everything started with *Ramapithecus*, a creature that lived at the margins of the tropical forests that were widespread across Africa some 15 to 10 million years ago. These creatures existed for a very long time, but eventually they turned into the oldest hominid populations called *Australopithecines*, discovered by the Leakeys at Olduvai Gorge in Ethiopia. *Australopithecines* also seems to have been fairly stable and it was only 3 to 2 Ma that we witnesses some morphological changes that suggest a shift toward modern humans, including increased brain size and a more upright posture.

For Keesing, these changes indicated increased intelligence that resulted from hunting, the use of tools, and life in organized social groups. These practices combined in a way that favored the evolution of intelligence. In the standard account, this happened because making tools involves a developed capacity to plan a sequence of steps; hunting requires the capacity to communicate plans, and living in campsites requires cooperation, all of which provided these creatures with a pattern that, once set in motion, placed more of a premium on intelligence.

This increased intelligence and the patterns of organized life that accompanied it "made it possible" for hominids "to cross another threshold about 1.3 Ma." These new creatures became known as *Homo erectus*, and over a million-year time span they evolved gradually into *Homo sapiens* (Keesing 1981: 11–16) with his fully developed ability to communicate symbolically. From then on, human biology was not seen as prior to culture but as expressed through culture (Keesing 1981: 28). And therein lies the miracle: suddenly humans had developed an open-ended system of behavioral potential that was expressed through locally developed symbolic systems. They could become creatures of all kinds of environments (Keesing 1981: 28).

It seems that this narrative of human evolution then became so trivial or well accepted, that it became *the* story of human evolution. It was simply accepted that humans, at a certain point, stepped out of nature, thenceforth to live by culture alone. This is presumably the reason why this chapter about evolution was subsequently removed when Andrew Strathern (Keesing and Strathern 1998) reedited the book after Keesing had died. The assumptions about evolution in Keesing's edition had seemingly become reality.

From then on, cultural holism and relativism seem to have become stable scientific truths in anthropology. But, both cultural holism and relativism owe us an explanation as to how a spontaneously symbolic differentiation could arise from a common technological base. And given that fiddling with tools is given evolutionary priority, and promoted as its major motor, it becomes quite difficult to understand this sudden need for diversification.

There is nothing in this evolutionary narrative that can account for the spontaneous emergence of the incredible cultural diversity that somehow materialized out of a common technological situation. This is to confuse ends and means. It is one thing to believe that early humans invented the sign, but quite another to assume that because it was a sign it was naturally also arbitrary. That is reading history backwards.

It should by now be apparent that the only "driving force" in this story is internal interaction between humans in the different stages of human evolution, which propels humans forward toward the next stage by its own internal momentum. Humans developed intelligence by tinkering with tools and living in social groups. This enabled humans to escape nature, and somehow also their identical mode of interaction produced different cultures and behaviors. But no other animal ever evolved in this fashion. If we are to believe Darwin, every single species evolved in relation to the environment.

ISSUES IN HUMAN EVOLUTION

It is after all, a rare sight to see animals walk on two legs to a marketplace, buy food from other animals, then stroll home to cook it and talk to their family about the shopping trip. There must be an evolutionary explanation for this eccentric behavior.

How did we end up like that? When did we utter our first word (Kenneally 2007: 1) so we could ask for the commodities we need? When did the dawn of human culture (Klein and Edgar 2002: 154) occur that made all this possible? What prehistory of the mind (Mithen 1996: 160) enables us to reflect about such things? Why do humans look so strange? And why do we act in such peculiar ways? All these much-debated questions revolve around the basic issue of what it is that makes us human (Pasternak 2007), and so very different from all other animals. Specifically: how and why did humans get culture? How did human evolution prepare humans for culture? And finally, what is culture?

Several important contributions have proffered answers to these questions about human evolution. It has been suggested that fire (e.g., Burton 2009), tool-making (e.g., Gibson and Ingold 1993), trade (e.g., Ofek 2001) or niche construction (e.g., Bickerton 2009), respectively, each played the essential role in human evolution. There has also been a prominent focus on the development of language, because it is through language that humans' symbolic capabilities are most vividly expressed. So we must first scrutinize these other explanations if we are to eliminate them as possible explanations for the final transition into our cultured *Homo sapiens*.

Even though in Evans-Pritchard's time-grappling with human customs and belief was considered mere speculation, times change, and new evidence of human evolution is now accumulating. Nowadays, for instance, the evolution of language is widely debated, and is no longer despised as a topic of study (Kenneally 2007: 7).

Mithen (1996), for instance, focuses on the prehistory of the mind. He finds that the selective pressure on humans to develop a stronger mind came about as a result of early human engagement in tool-making (1996: 108), interaction with the environment to find prey (1996: 116), and life in progressively larger groups, which eventually promoted the evolutionary selection of social intelligence (1996: 118). These conditions created talkative individuals who gained selective advantages by exploiting the nonsocial knowledge of their not-so-gifted peers (1996: 213). In a sense, then, early humans talked themselves into becoming *sapiens*.

Burling (2005) has also studied the evolution of language. For him, language arose as a result of a need for communication. Thus, "selection for better communication brought language into existence and fostered its development" (2005: 91). In Burling's account, the need to communicate simply intensified at some point in prehistory, turning an animal communication system into articulate human talk. But he does not address the question of why the human population selected for language in the first place.

Bickerton tries to answer this question. For him, language is simply what makes us human (2009: 4). It was this linguistic ability that triggered the evolution of the human mind (2009: 9). But contrary to Burling, he finds that different species' communication systems are normally only designed to take care of that particular species' evolutionary needs, and hence there is no evidence that communication in general becomes more sophisticated over time (2009: 13). Instead, he argues that there must have been a radical change in our ancestors' behavior that enabled them to break through the limitations that restrict all other animal communication systems.

Bickerton relies on niche construction theory, which claims that species change their environment, and that this will subsequently select for new traits in these species (Bickerton 2009: 11). According to this theory, the human species broadened its capacities precisely through such successive niche constructions (2009: 154). This went on until one day the "species–niche" continuum had reached a point where a breakthrough was possible. This reportedly happened as a result of a new need to recruit help for common tasks like bringing home a particularly large game animal. This need had to be communicated to the larger group, or they would go hungry—and so they began to talk to each other (2009: 164). But, the archeological records do not seem to endorse Brickerton's scenario, since we know now that niche building was not a continuous process but based on long intervals of stasis and more abrupt periods of change.

There are also theories about external phenomena that are believed to have had a great influence on human evolution. A key candidate is fire. Burton (2009) thinks that fire was the spark that ignited human evolution, and that the very act of associating with fire was the turning point in human evolution (2009: 4). He argues that early *homo* species had to live with fire because the African savanna historically underwent widespread burning as part of its natural reproduction cycle (2009: 20). A similar argument is made by Pyne (2001: 183), who rightly observes that all human societies have fire, all humans manipulate fire, and only humans do so (2001: 24). And eventually humans learned to domesticate fire and put it to their own uses.

Certainly, a great deal of human life is based upon the domestication of fire. As Wrangham (2009) argues, fire makes it possible for humans to cook their food, and cooking increases the amount of energy that the human body can extract from its food (2009: 14). But as far as we know, fire has only been domesticated since *Homo ergaster* and his relatives. So fire cannot be the whole story.

Others claim that the ability to trade underpins human evolution. Ofek (2001; see also Saint-Paul 2007), for instance, notes that the propensity to exchange between two unrelated persons "is a profound distinguishing feature of human existence" (2001: 1). Inventing trade must have been a turning point in human evolution, because as Ofek also points out, all contemporary human societies literally live and die by exchange (2001: 23). Yet while trade might well have been a turning point, we only find evidence of exchange after the advent of *Homo sapiens*—so trade cannot solve the whole mystery either.

Still others focus on human cultural evolution. Two main schools frame this debate: Darwinian anthropology and evolutionary psychology. Darwinian anthropology has focused on the functional consequences of behavior and their immediate returns in terms of energy acquired and spent, or their long-term fitness in the form of genetic representation in future generations (Dunbar, Knight, and Power 1999: 2). Evolutionary psychology, on the other hand, has focused on the cognitive mechanisms believed to underpin behavior. These scholars argue for an environment of evolutionary adaptedness that spans two or three million years during which the distinctive human mind evolved in adaptation to life as a prehistoric hunter-gatherer (Dunbar, Knight, and Power 1999: 2).

All these explanations of human evolution assume the continuous development of abilities and cognitive skills or a kind of cumulative niche construction that at some point enabled our human ancestors to create a symbolic universe. Unfortunately, the fossil record does not seem to corroborate this.

Gibson (1993) further points out that tool-making and language are not all that particular to the human species. She argues that field and laboratory studies of animal behavior have shown that there appear to be few behavioral discontinuities between apes and humans. She lists a number of studies that suggest that several chimpanzee and gorilla species, as well as other animals, have symbolic capabilities, can use syntax, and are able to use tools to make tools (Gibson 1993: 6–7).

These abilities were and still are important factors in human lives—humans everywhere use tools, make fire, and trade—but the archeological evidence does not seem to substantiate any of these as the defining traits on the human path to culture. Rather, each of these abilities can be attributed to early hominines at different evolutionary stages (Tattersall 2012). And each of these human abilities corresponds in evolutionary terms to an enlargement of the exploitable environment. Walking made it possible to adapt to the savannah and the resources found there. Fire extended the exploitable environment further by making more resources edible. So, if *Homo sapiens* is not a result of a freak gene mutation we still need to find the adaptive mechanism that both expanded the niches that humans were able to exploit, and which also contained the seed of symbolism. And as we shall see, that mechanism is kinship.

HUMAN EVOLUTIONARY SEQUENCE

There is a certain consensus about the sequence of human evolution. The Committee on the Earth System gives the following sequence:

> 1. The split from chimpanzees at 8–6 million years ago (Ma). 2. The development of bipedal locomotion, probably occurring at this split. 3. The slow evolutionary change to bigger teeth, thicker enamel, and reduction of canines that characterize a 5-Ma-long lineage from *sahelanthropus* and *orrorin* (if those are not the same as *Ardipithecus*), through *Ardipithecus*, to *Australopithecus*, and finally to *Paranthropus*. 4. A splitting event between 3 and 2.5 Ma that produced *Homo* from an *Australopithecus* ancestor. 5. The development of stone tool technology at about 2.6 Ma. 6. The origin of a more carnivorous species, *Homo erectus*, at about 1.9 Ma. 7. The first dispersal by hominins out of Africa, by 1.8 Ma. 8. The development of the Acheulean stone tool culture at about 1.6 Ma. 9. The origin of *Homo sapiens* at about 200 ka, and then out of nowhere, 10. The origin of symbolic language. (2010: 22)

However, if one relates the known phases of human evolution to what have, for a long time been considered its prime movers (fire, cooking, walking upright), it becomes clear that no plausible adaptive explanations are given for any of these phases, and certainly not for the last transition into *Homo sapiens*. So we need to take a critical look at this sequence of phases. First, the above discussion makes it abundantly clear that we must relinquish the illusion of "man the tool-maker," and "man the hunter." We must slay our heroes. These characteristics were part of the human tool-kit at different times, but they did not singularly drive human evolution and they were certainly not the driving force behind the later dawn of culture.

It is also evident that these monocausal explanations need to be put into the right chronological order. To do so we still need a prime mover, and as always in the animal kingdom that is the environment. The human evolutionary sequence as we know it resulted from the challenges humans faced in garnering sufficient resources in a continually changing environment. Different eras called for different solutions to severe climatic shifts. Each successive *homo* species was just trying to survive. But by adjusting their behavior to the new conditions, humans simultaneously extended their ecological niches. At each stage, behavioral adaptations arose as a response to stress, but when times improved these adaptations provided them with immense, cumulative advantages from each evolutionary stage to the next.

There was nothing inevitable or sudden about this evolution. It resulted from a sequence of selective and unconnected adaptive transformations that occurred in response to changes in the Earth's climate, and it took several millennia. There was nothing predestined about it, no inherent potential that just *had* to unfold, no heroes and no march to glory. Just evolution doing its thing over time. Ultimately, humans ended up doing exactly what evolution dictated. They never did escape from nature.

But we are still left with some prickly issues concerning the emergence of the human lineage. One of the most important is how and why humans started to walk upright on two legs, and why humans live in large groups. One of the more convincing explanations for these two human characteristics is given by Hart and Sussman in their intriguing book *Man the Hunted* (2005). In their book they turn the whole early human evolution story upside-down to arrive at a much more plausible explanation for why, how, and when early humans became bipedal and started to live in larger social groups.

The powerful images of man the hunter and man the tool-maker have long populated our imaginations with a cast of colorful characters: valiant hunters and brave fighters but also their alter egos who spread war and genocide. Seen from this perspective, humans were thought to have evolved from some kind of belligerent killer-ape resembling the *Neanderthal*, who lived a rough life slaying large beasts in a hostile environment. Had *Neanderthal* really been our forefathers much violence could have been explained away by our inherited characteristics. Violence would simply have been in our genes.

That is partly why it is so important to get the story of our human origins right, and to correct imagery and explanations that muddy our understanding of the present human condition. We know for sure, now, that our ancestors were not the *Neanderthal*, nor some other form of killer-ape. Instead, Hart and Sussman suggest a more ignoble beginning, or at least a more humble one. They posit that early humans were bipedal edibles, tasty primates that were hunted and eaten by a whole host of predators (2005: xvi).

Based on new archeological finds, Hart and Sussman recast our understanding of some fundamental issues in early human development. In 2001 a French team lead by Michel Brunet made a find in Chad consisting of a cranium, a jaw fragment, and several teeth. They called it *Toumai*, or "the hope of life" in the local language, also known as *Sahelanthropus tchadeniss*. This find stretches the human lineage back in history to between 6 and 7 million years ago and it seems to refute existing theories concerning the location and habitat of hominid origins.

Until this find, all hominid fossils had been found on the east side of the Rift Valley, and this situation had led to the belief that when chimps and hominids evolved from their common ancestors the chimps line stayed in the trees of the forested region to the west of the Rift Valley, while the hominine began to colonize the drier savannahs of the east, thereby developing bipedalism (Hart and Sussman 2005: 14-5). But the *Toumai* find on the "wrong" side of the Rift Valley spoils this theory. There seems to be no obvious link between wandering out onto the open savannah and evolving bipedalism, as was traditionally assumed.

Toumai lived long before the vast growth of open grassland in Africa, long before the increase in size of the human brain, and long before these creatures began to make any recognizable stone tools. And the toes of the skeletons found also indicated that something else may have been at stake. The 3.2 million year-old skeleton of Lucy found by the Leakeys is bipedal but had long feet with a very strong big toe that was divergent like the modern human thumb, and which could be used to grasp and climb trees (Hart and Sussman 2005: 17). Because of this toe, these hominids were able to live both in trees and on the ground. So for a long time walking on two legs was not the same matter as it is today. Interestingly, this thumbed foot predates human arrival on the savannah, making it unlikely that bipedalism was as an adaptation to life on the savannah.

So far there have been six different explanations for bipedalism: the carrying model, the vigilance model, the heat-dissipation model, the energy-efficiency model, the display model, and the foraging model. None are entirely convincing. *Carrying*: Bipedalism might have released human hands for other duties but the anatomical features necessary for carrying are not present in early human species. *Vigilance*: Standing upright may have enabled a better overview of the immediate environment—but many other mammals also stand in order to scan the environment and none of them have ever learned to walk on two legs. *Heat-dissipation*: Standing upright means less surface area exposed to direct sunlight which might be a plausible explanation if early hominids had evolved on the savannah. However, current theory holds that they evolved in open woodland and on the edges of forests where the sun plays a lesser role. *Energy-efficiency*: Less energy is consumed when walking upright, but it seems that the anatomical changes needed to do so first occur *after* the onset of two-legged locomotion. *Display*: Standing upright makes humans look larger and more imposing than they really are and can be used to display dominance, but chimpanzees and other animals that cannot walk also use this tactic. *Foraging*: Standing on two legs may have helped humans to reach otherwise unattainable fruits and other edible things, but it is difficult to see how reaching up instigated walking (Hart and Sussman 2005: 175–76).

Each of these factors provides some advantages to living on the open savannah, and they also all involve some adaptation. But none of these adaptations seem to be part of a general evolutionary thrust affecting all animals on the savannah, all of whom do just fine without them in the same circumstances.

In lieu of all these theories, Hart and Sussman put forward a more convincing explanation for the unique upright posture of humans. They propose that the early humans who came down from the trees already had an upright posture. If this is so, then we are not dealing with a radical change in behavior, which, seemingly out of nowhere, changed quadrupedal animals into bipedal ones. And, as Hart and Sussman note, there are a number of primate species—the vertical clingers and leapers—whose posture is upright both in trees and on the ground but who are unable to walk quadrupedally. Instead, they hop bipedally in an upright position (2005: 176–7). These monkey species did not have to learn to rise from all four when they came down from the trees, and this is also likely the case for early hominids.

These primates were pre-adapted for bipedal locomotion—the question is what forced them to the ground. The answer seems to lie in a mosaic of factors in a changing environment, not least climate change (Hart and Sussman 2005: 178). Early hominids lived at the edge of the forests and when the climate became dryer and cooler a wide range of edible ground-level plants appeared.

Bipedalism would only have been advantageous if early hominids had left the trees when the majority of survival activities took place on the ground. In such a transition, all the above mentioned factors would have benefitted early bipeds living on the open savannah, and they might together have helped select for continued two-legged locomotion. As Hart and Sussman state: "there is less chance of being molested by predators if you incorporate scanning, looking larger, walking quickly and efficiently, carrying food back to a safe place, and carrying an infant by moving on two legs instead of four" (2005: 178).

Early hominids were prey animals, but they became adept at deterring predation through alarm vocalization, group living, constant visual scanning of the environment, throwing things, and intimidating attackers by standing up (Hart and Sussman 2005: 178). But there are still no signs of the mythical killer-ape.

Neither were these creatures hunters. They simply did not have the dental equipment for meat-eating. Hunting is only a good survival strategy if you are able to eat meat, and to eat meat you have to be able to chew meat (Hart and Sussman 2005: 226). Or else, you have to be able to cook it, and we know that the domestication of fire came much later. Early hominids also were not great tool-makers. They may have used the odd stick or stone, but there is no archeological record of this. Proper tools came later.

Early humans were just vulnerable mammals in a hostile environment. To survive, they had to stick together. They were social beings who lived in groups and who, early on, became adept at vocalization. Those were the characteristics that defined what it meant to be human, and it was these characteristics that evolution had to build on in order to produce a new species in the human lineage. In any case, these original walking ape-men existed for the better part of 3 million years.

The next human-like change entered the scene with *Homo habilis*. This seems to be the first species that used a proper tool. That tool was simply a stone with rough flakes hacked off it. It has been termed the Oldowan complex, after the Olduvai Gorge in Eastern Africa where it was first found. Yet this tool-making does not appear to have enhanced *Homo habilis*'s cognitive and inventive power—in fact *habilis* unimaginatively used the same tool-kit for around 800,000 years (Wade 2006: 18–19). His tools may have been a sign of some basic form of intelligence, but they did not do much to enhance it.

But suddenly, after having used the same tool for almost a million years, some of these bipedal apes finally had a bright idea. They invented the so-called "hand-axe." This consisted of a biface stone that was flaked to a greater or lesser degree on both surfaces (Klein and Edgar 2002: 102). This is called the Acheulean tool complex (Klein and Edgar 2002: 105). This new technological complex was also remarkably enduring, however. From its inception roughly 1.65 million years ago until its disappearance about 250,000 years ago, it remained almost unchanged (Klein and Edgar 2002:108). It took the emergence of *Homo sapiens* to change all this.

So, rather than a steady increase in capabilities and technological know-how, we see long periods of stasis interspersed with the relatively sudden and very sporadic appearance of the odd major novelty (Tattersall 2007: 137, in Pasternak). This lends little credit to explanations such as the tool/language/symbolic behavior sequence, the domestication of fire, or the continuous niche construction as the sole factors in human evolution.

The different technological complexes therefore do not help to explain the shifts in human speciation. In this regard, the Committee on the Earth System Context for Hominin and National Research notes that the Acheulean tool complex normally thought to belong to *Homo ergaster* and *erectus* had not yet been developed at the time their first fossils occurred (Committee on the Earth System Context for Hominin and National Research 2010: 21). Oppenheimer tellingly points out that although modern humans have been around for at least 160,000 years, for around the first 100,000 years they made and used the same tools as their predecessors (2003: 98). Mithen also notes that the new *homo* used the same range of tools as his forbearers for quite some time, but he also noticed something new: the most recent *homo* species buried their dead and used lumps of red ochre for coloring objects and perhaps also themselves (1996: 20).

It therefore seems unlikely that progressive advances in technological knowhow were the motor of human evolution, so we can now definitively dispense with the tool-making hypothesis and all its derivatives. Instead, we need to look at other kinds of changed behavior that are independent of technology, notably the different *homo* species' interactions with the environment and the resources they were able to exploit. All living things have to interact with the earth's environmental system, which consists of land, oceans, and atmosphere. The study of human evolution shows that, like other organisms, humans have evolved over a long period of time in response to environmental challenges and opportunities (Committee on the Earth System Context for Hominin and National Research 2010: 10).

ON TWO LEGS ON THE OPEN SAVANNAH

We know now that human evolution was not propelled forward by tinkering with tools, and that humans seemingly ventured out on the savannah already adapted to bipedalism. Now we need to sort out how and when the other presumed prime-movers in human evolution emerged and to relate these to changes in environmental conditions. To begin exploring the relationship between human evolution and the environment, the Committee on the Earth System observes that human evolution is closely related to the development of the East African Rift System and its flanking mountains.

This system regulated the extent of aridity and occurrences of water resources that early hominines relied upon in various parts of tropical Africa (Committee on the Earth System Context for Hominin and National Research 2010: 23). The Rift System also regulated the extension of the African savannah environment which responds primarily to the amount and seasonal distribution of rainfall. We know that fluctuations in the savannah system shaped human evolution.

There is, then, good reason to believe that humans evolved in response to severe climate changes. As the Committee on the Earth System notes, there is now evidence that several major junctures in human evolution and behavior coincided with fundamental changes in global and regional climate.

The climate record of sub-Saharan Africa shows that with some regional exceptions Africa responds to glacial periods with aridity that results in expanding deserts and arid lands, while interglacial periods result in a wetter climate and more woodland (Marean 2010: 429). Cold periods would then be times when African populations experienced bottleneck and split catering for speciation, while warm intervals would give periods of population expansion and the absence of splitting events (Marean 2010: 430). That leaves us the task of sorting out the correlation between the different attributes that humans developed in these changing climate conditions, figuring out the timeframe, and considering the consequences for human evolution.

Early ape ancestors in the human lineage are associated with a general global cooling from 8.5 to 6.5 Ma. At this time, early ape ancestors to the human descent line, *Orrorin* and *Ardipithecus*, probably lived in predominantly closed woodlands or forests. *Aridipithecus ramidus* lived in environments that varied over time or favored areas that contained both forest and open habitat, while *A. anamensis* lived in relatively open habitats (Committee on the Earth System Context for Hominin and National Research 2010: 32).

This situation lasted until the onset of the major Northern Hemisphere glaciations 3 Ma, when there was a progressive replacement of closed forest woodland by open savannah grassland. Around 2.3 Ma, a faunal turnover occurred, with an almost complete replacement of mammalian *taxa* at many localities. Some of the early hominine species found themselves living in an increasingly treeless landscape, and with the advent of the Northern Hemisphere glaciations at 2.8 Ma some of them were living on the open savannah, faced with a rapidly changing fauna and animal composition. Those who found themselves living in these surroundings had to make the best of diminishing vegetable resources (Oppenheimer 2003: 14). And since it is quite difficult for large mammals to live by foraging alone because plants and insects do not provide enough nourishment, they had to include meat in their diet (Wells 2002: 89).

This period corresponds to the appearance of a new hominine species, *Homo habilis*, which was the earliest *homo* species (Committee on the Earth System Context for Hominin and National Research 2010: 34). These early humans probably obtained most of their meat by scavenging and occasional hunting, with some protein possibly also obtained from insects. The African savannah undergoes widespread seasonal burning and many birds and mammals congregate there to garner the resultant food resources such as seeds and nuts, insects, beetles, and small vertebrates which lose their cover in the burned environment, exposing them for the picking (Burton 2009: 21). In this way, *Homo habilis* was accustomed to recurring fires, and might have homed in on burning areas to await dinner (Burton 2009: 176). These early human might also have discovered that patch burning promotes the growth of berries, fruits, or flowers and that fire made the landscape more accessible (Pyne 2001: 51).

In these times of dwindling resources, early humans' dietary resources were less densely distributed and they would have been forced to become more efficient at exploiting them, for instance by moving around (Lewin and Foley 2004: 250). Any such movement would have had to be cost effective. This may have helped to accentuate and promote bipedalism. As Cela-Conde and Ayala notice, bipedalism did not initially have to be full blown, as Hart and Sussman also suggest, but could have consisted of a kind of slow bipedalism with an excellent energetic balance in low-velocity foraging tasks (2007:103).

However, having located the carcasses and driven other predators away, the early humans still had a major problem, namely all the tasty meat and marrow that they could not get at. Human teeth were not up to this task. They needed tools. So they used sharpened or flaked stones, bringing the Oldowan technological complex into existence. In this way, they increased their ability to procure food from the environment. These adaptive changes made it possible for them to take over the savannah. They may even, at that time, have been the true kings of the savannah.

RUN FOR YOUR LIFE

Homo habilis seem to have been very much at home in their new environment, and not a lot happened for another 800,000 years. But nothing lasts forever. Eventually, another series of ice ages dried up much of Africa, and large areas of the savannah turned into grassland and changed the composition of the animals living in the area (Committee on the Earth System Context for Hominin and National Research 2010: 35; Oppenheimer 2003: 15). The period from 1.8 Ma until 1.0 Ma saw the greatest expansion of grassland vegetation in Africa to date, and thus many early human localities consisted of fairly open grassland habitats. During this period, many types of fauna in Africa were replaced by grazing animals (Committee on the Earth System Context for Hominin and National Research 2010: 35).

One consequence of all this was that there was no longer enough food to go around. This new situation called for a creative solution. *Habilis* had been accustomed to eating meat. But hunting requires a great deal of energy: "it's what you do if there's no reasonably fresh meat already lying around" (Bickerton 2009: 119). The increased proportion of grazing animals may also have meant that there were fewer smaller carcasses lying around. *Habilis* could no longer live by scavenging alone: they needed whole animals to be able to survive. Their stomachs were empty and calling for action.

It was also at this time that we see a significant change in the relationship between animals and humans in terms of who got to the prey first. Up until this time, humans had always been among the last to arrive on the scene. Cut marks show that human tools were used only after the carcasses had been eaten by animals. But suddenly the sequence of marks was reversed (Bickerton 2009: 125). Humans came first, suggesting that they had become hunters themselves. This was probably the first hunter-gatherer society.

At that time, some major changes also occurred in human morphology. For instance, the relative relationship between legs and body mass increased by up to 50 percent in comparison with *Australopithecus afarensis*, but we also see larger hind limbs and vertebral joint surfaces, narrower waists, and shorter toes. These features have traditionally been interpreted as a result of adaptation to improved long-distance walking in more arid and open habitats. But there are other features that are not solely connected to improved walking. Bramble and Lieberman suggest that these specialized features are tightly connected to running. They include:

> An extensive system of springs in the leg and foot the effectively store and release elastic energy during running; hypertrophied gluteus maximus and spinal extensor muscles that contract strongly to stabilize the trunk in running but not walking; and an elongate, narrow waist in combination with low, wide, decoupled shoulder girdle that have an essential stabilizing function only in running (2004: 351)

We can see from the archeological evidence and current morphology that humans adapted to become good long-distance runners. This caused some special bodily adaptations: humans developed long tendons, big buttocks, a balanced head, short toes, a narrow pelvis, and stable foot arches (Stone, Lurquin, and Cavalli-Sforza 2007: 27).

All these features make it possible for early humans to perform endurance running, which no other primates are capable of. These new *humans* seem to have evolved to travel long distances by both walking and running, in accordance with their changing environment. Long distance running also depends on the runners' ability to sweat, not just through their mouths but also through their skin. Fur is a disadvantage here. So we may conjecture that running was the reason humans lost their fur. This relatively hairless breed of runners was to become known as *Homo ergaster*. These humans were "born to run." They had to change their behavior to cope with a new situation: the acute need for food. So they changed their diet, leading to the unique contemporary human combination of large bodies, small guts, big brains, and small teeth (Bramble and Lieberman 2004: 351).

It has been suggested that endurance running evolved for predator pursuit and may have helped hunters get close enough to throw projectiles, or perhaps even to run some mammals down by heat exhaustion. It may also have been useful for scavenging in the open, semi-arid environment by helping early humans to compete more effectively for the scattered resources for which they might have had to run long distances to compete with other scavengers.

But, perhaps, they just started out by copying what they saw other hunting animals do. They might have run toward a grazing flock to separate one animal from the flock, and then tried to run it down. Whereupon it would outrun them. But these human hunters could then *stand up* and watch the animal flee into the distance. Sooner or later, it would run out of steam and lie down. If the humans were hungry enough, they might have given chase again. And the scenario would repeat itself. But needs must be met when the devil drives, so they would have had no choice but to resume the chase. Only by then their prey would have been exhausted and unable to run any further. The faster *habilis* could run toward the escaping animal, the quicker the catch. That means that the fastest hunters brought their group more meat, and were better at surviving, than groups with poor runners, who would quickly have become extinct. The survivors became known as *Homo ergaster*.

There was another problem too. Irrespective of whether they were just scavenging or actually hunting, the landscape was changing. Their savannah was swiftly being replaced with grassland (called C4 grass). Grassland is not exactly open landscape. Lightning is, as Burton points out, the primary ignition source that maintains the savannah system (Burton 2009: 20), but with the drier climate came longer spells without lightning and thus also without fire, which meant that the grass on the savannah grew dense and tall.

To understand the early human response to the ever-denser grassland we have to look at the ethnographic material from the Aborigines in Australia (Elton 2008). As we now know, Australian Aborigines were the direct descendants of early *Homo sapiens*, and therefore they probably shared the same skills they had inherited from the early *homo* lineage. So we need to look for behaviors that made it possible for the Aborigines to exploit a landscape dominated by dense grassland.

In Australia, this same type of grassland seriously limited the Aborigines' movement. In 1923, Stuart landed in Vansittart Bay, Australia, where he encountered twelve-foot (3.6 m) high grass. Similarly, Love wrote in 1936 that huge areas of the place were choked beneath grass up to eight feet (2.4 m) tall, which dried out fast (Vigilante 2001: 146, 147). This would have seriously limited hunting, which in the African case was limited to a very short period at the onset of a new rainy season when the lightning returned.

Early *Homo ergaster* had to imitate nature. The solution was to burn the grass. In fact, we know that the Australian Aboriginals habitually burned the land through which they passed. As Stuart continues: "In most places the natives keep this rank growth burnt off along the coast to keep the country more accessible" (Stuart in Vigilante 2001: 146). One single wandering band could ignite 5,000 fires a year, just to control the environment (Pyne 2001: 54). Had they not done this, they would not have been able to move around, and their game would have been almost impossible to track down. We can thus posit that it was the overgrowth of this particular type of grass that constituted the selective pressure which forced *homo* (by now *ergaster*) into domesticating fire. It may well have taken them a thousand years of experimentation to do it right, but by archeological standards this is but the blink of an eye. We know from later evidence that they did exactly this. From sites such as South Africa's Klasies River Mouth Cave, humans deliberately burned grassland more than 60,000 years ago. They did this to encourage the growth of nutritious tubers, which germinate after exposure to fire (Wong 2005 Internet version p. 4).

Domesticating fire had other positive side effects. *Homo* was now hunting down whole animals that had to be butchered, and the large hunks of meat had to be made edible. Fire helped them to digest their food (Mann 2000; Wrangham 2009), and as a result of this discovery, many more things became instantly edible (Leonard and W. R. L 2003; Mann 2000). Suddenly, *Homo ergaster* had extended his biological niche. He could inhabit a larger habitat; as long as there was something there to cook. Cooking also released more calories for long distance running. So we might think of *ergaster* as a kind of "running cook." In this way, *ergaster* initiated the transition from the raw to the cooked (see Lévi-Strauss 1979).

Sometime later, the weather got better and *Homo ergaster* began to colonize new territories that had been hostile to him before his newfound cooking abilities. Now he could survive as long as he could find something to cook within his foraging range. This innovation was such a dietary success that life without cooking is practically unthinkable today.

But things were not entirely straightforward. The new hunting habits posed new problems. *Ergaster* were still slicing away at their game with their well-tried stone tools. These might have served them well to slice the last bits of meat off scavenged bones, but they were not up to a complete butchering process, as the edges quickly became dull. This explains why *Ergaster* made tools on the spot. They took stone nodules with them when hunting, and processed them around the carcasses.

As Jones notes, in the hours that followed, the edges on this tool would be among the sharpest ever fashioned by man (Jones 2007: 46). New behavior produced the incentive to invent a new tool, leading to the Acheulean technology complex which we know today as the fossil trademark of *Homo ergaster* and the *Neanderthal*. This was a byproduct of *ergaster's* interaction with a changing environment which made two-legged running, fire, and cooking part of the human condition ever after.

TOWARD KINSHIP

By now pre-*sapiens* populations were able to hunt, make fire, and cook their food, but they were still not *Homo sapiens*. They had also laid the basic foundation for the next evolutionary step: they had become more sociable. These new abilities must also have induced *Homo ergaster* to engage in a more communal life.

It seems reasonable to posit that *ergaster* developed some kind of in-group solidarity through sitting around the camp-fire eating together—maybe even an incipient notion of kinship. Even among the nonhuman primates there is evidence of kin recognition, and social organization is influenced by distinct patterns of dispersal of the young when they reach maturity. Chimpanzee females, for instance, typically leave their group at maturity and move to live with another group (Stone, Lurquin, and Cavalli-Sforza 2007: 218, 219).

Young female *ergasters* probably also migrated to other groups consisting of consanguine males. Their communal life is likely to have enhanced the notion of kinship (see Chapais 2008). According to Wade, the reduced body dimorphism between *ergaster* males and females suggests some important change in social structure that also involved a switch from the separate male and female hierarchies of chimp communities to a male-female bond that could mark the beginning of a family structure (Wade 2006: 21).

We find another indicator of this in anatomy. *Ergaster* had a narrower pelvis and thus a smaller birth canal, which meant that much of the infant's extra brain size had to have been acquired after birth, and infants therefore would have had to be carried in arms for longer periods while their brains finished developing, which according to Wade would have resulted in a closer bond between parents (Wade 2006: 22). There was obviously a greater investment in offspring. These groups were linked to each other through their female offspring. And with the strengthening of kinship bonds, this link probably became more prominent. This is an important issue in our understanding of the later changes that *Homo sapiens* underwent. So, at this stage we can already discern some of the fundamental anthropological and ethnographic staples, such as patrilineal bands and exogamy. But each group still consisted of a single foraging group that had access to all the necessary resources within its territory.

In sum, *ergaster* consisted of self-sufficient groups that could obtain their provisions from their immediate environment. So they roamed the landscape in parallel groups that were organized in the same way, ate in the same manner—and cooked everything that could be cooked in their vicinity. This organization made it possible for them to colonize large parts of Euro-Asia. They were on the brink of becoming *sapiens*.

Evolution need not have taken this route to intelligence. In fact there was no reason at all for this to occur. One ice age too few or too many, or more or less severe, could have changed the whole story. Still, it is unlikely that any other species would have followed the same evolutionary path. It all happened in a series of haphazard circumstances that only had this particular effect on humans—in other words, it was not just part of a general evolutionary thrust toward intelligence that birds or ants, for instance, could have won.

One thing is for sure, human evolution is not a hero story. Nor is it about intelligence slowly unfolding. Humans did not evolve from some kind of killer ape whose influence can still be felt. Humans evolved through a long process of spurts and periods of stasis in connection with a changing environment, and the spurts were mainly connected to recurrent ice ages that changed the conditions of the *homo* populations.

There was nothing inevitable about this particular development trajectory. But the *homo* lineage was pre-adapted to two-legged life on the savannah when these changes occurred as a result of one of many ice ages. And life on the savannah prepared them for more changes when yet another ice age changed the conditions on the savannah once more, posing new challenges which the early humans coped with through greater mobility and by producing fire, which enabled them to cook their food. They became sociable, vocalized, mobile, and able to procure food from a large variety of habitats. They were able to colonize the better part of the globe. In fact, they were just waiting for the next ice age.

Chapter Two

Connecting Niches by Kinship

Homo ergaster had many skills. They were sociable. They hunted and gathered in groups based on sex. They sat around the camp fire in the evenings eating their cooked food. But they still lived in separate groups who foraged on their own, exploiting their territory to the best of their abilities. They still needed to find all their vital resources within this area. In contrast, we know that *Homo sapiens* exchange products between groups. We also know that *Homo sapiens* began to exchange things less than 200,000 years ago. What caused *ergaster* to adopt this strategy?

Genetic studies reveal that an African woman from around that time was our common ancestor. In the media she became known as *Eve*. Modern humans arose in one place and spread elsewhere. Genetics tells us this, but it does not tell us anything about under what selective pressures this happened—that is the business of archeology—(Cann and Wilson 2003: 421), and neither of these disciplines suggest that a new behavior could have caused speciation.

Still, these two sciences concerned with human evolution are hard to reconcile because genetics studies genes, which always have ancestors; while archeologists study dead fossils that may not have any descendants. So we need to qualify the finds of both with the practices that these early humans instigated and passed on to subsequent generations. We can do this using ethnographic material and anthropological theory.

These new practices must have evolved as an evolutionary response to changed living conditions; practices that forever became part of the human condition. We also know that it was the descendants of this woman who "invented" culture. But while it is unlikely that this new breed of humans had culture from day one, they must still have undergone a behavioral change that required nascent symbolism. And this must have resulted from another practice that was new to *ergaster*.

Somehow, the life conditions of *ergaster* must have been challenged by changes in their environment. *Ergaster* had been very complacent in their niche. Their technological complex lasted for more than a million years. What changed their situation seems to have been the advent of one of the coldest ice ages ever, which began a half a million years ago, making life on the savannah grassland particularly harsh due to a prolonged period of mega-drought and a resulting dramatic shortage of resources (Cela-Conde and Ayala 2007: 329).

According to Wells, this ice age turned into a deep-freeze: the climate deteriorated, the forests shrank, and life became more difficult (2002: 95). This was the era when the first *Homo sapiens* appeared. This new breed appeared around 200,000—170,000 years ago, seemingly born out of an almost total human extinction resulting from what Oppenheimer calls "the mother of all ice ages" (2003: 16). This caused tropical and subtropical mega-droughts that would have brought about desertification on the savannah grassland system, making it extremely inhospitable for its inhabitants. *Homo ergaster's* ability to cook was no longer enough to enable survival in the desiccated environment, as there was no longer much left to cook. Recent genetic studies estimate that the population was, at this time, reduced to a mere 10,000 or even only 4,000-5,000 individuals (Wells 2002: 58), or perhaps even fewer, as Oppenheimer suggest (Oppenheimer 2003: 46). The *Homo* lineage had become an endangered species.

This ice age was by far the toughest, it was both the longest and coldest and it lasted from roughly 195 to 123 ka, causing expanded deserts and arid lands which cut off possible avenues for the movement of individuals and groups, and also blocked the road out of Africa. Therefore, human groups were stuck in small pockets spread around the African continent living under rather severe environmental stress for a significant period of time, all of which created excellent conditions for rapid evolution and population splitting (Marean 2010: 429). It seems reasonable to believe that the ice age isolated fragments of the *ergaster* population and that some of these groups experienced a bottleneck which reduced their numbers and caused speciation.

This situation helps to explain the fossils classified as *Homo sapiens* found in Ethiopia and North Africa, but which, according to Marean, appear to be quite far out variations of *Homo sapiens*, and are comparatively physically robust, so they may not be the lineage identified in the generic records purported to lead to modern humans (Marean 2010: 428).

This ice age practically wiped out the game which the human population had depended on (Oppenheimer 2003: 124), and it also limited access both to water and other easy prey, which might have lead some parts of the population to move to the coast (Wade 2006: 95). There are now strong indicators that those who fled to the coast may have been the only survivors, and therefore also our common ancestors. Curtis W. Marean, for instance, claims that the sea saved humanity (Marean 2010).

So, Marean started to look for a suitable site that could have given these early humans shelter and a means of survival during those harsh times. That site had to be close to the ancient coastline and elevated enough to preserve archeological deposits. With his colleague Peter Nilssen, Marean started to investigate some caves that Nilssen had spotted at a place called Pinnacle Point on the coast toward the Indian Coast in South Africa.

Here, they stumbled across a particular promising cave called PP13B. As Marean writes, "by all appearances, we had hit the jackpot" (Marean 2010 Internet version p. 3). Archeologists sometimes get lucky. The team then commissioned a farmer to build a 180-step staircase to allow safer access to the site, and they began to dig. The excavation took place between 2000 and 2006, and produced a remarkable record of inhabitation from between approximately 164,000 to around 35,000 years ago, which showed evidence of considerable behavioral sophistication.

Early inhabitation of PP13B coincides with the onset of the last ice age. And given their modernity they might just be the result of a speciation, or even early exemplars of the species in transition toward the new *Homo sapiens*. There is precious little archeological evidence from the period in which modern humans emerged, but based on fossil records this is generally believed to have occurred between 200 and 100 ka. The lack of evidence has made research into what actually happened rather elusive, but this is probably because the early population would have been rather small.

However, PP13B enriched this sparse body of evidence. Its finds document adaptation to coastal resources, the use of ochre, bladelet production, and complex technologies such as heat treatment (Marean 2010: 430). Marean even claims that this location on the southern coast of the Cape Floral Region could be the location of the progenitor population that emerged during the ice age, and that his excavation helps clarify their mobility system at the coastline and Cape Floral intersection.

This intersection forms a very diverse and complex ecosystem and there is evidence that it existed prior to 120 ka. In this region, we find the Cape Floral Region running along the western and southern shores of Southern Africa. This ecosystem hosts a wider diversity of plants and marine animals than any other region in the world. Based on its immense floral diversity it has been classified as its own floral kingdom, containing within it the Fynbos Biome that provides carbohydrates from its many plants with underground tubers and bulks. The Cape Floral Kingdom is the smallest of the six Floral Kingdoms in the world, and is the only one located entirely within a single country. It is characterized by its immense richness in plant species (8,700 species) and its high degree of endemicity (68 percent of plant species are confined to the Cape Floral Kingdom). The Cape Floral Kingdom thus compares with some of the richest floras worldwide, surpassing many tropical forest regions in its floral diversity (Marean 2010: 427).

The tubers that grow in this region have always been important ingredients in many hunter-gather societies due to their high carbohydrate yield; they are seasonally predictable and humans do not face much competition over these tubers compared to above ground fruits, nuts, and seeds. Furthermore, these plants (geophytes) are adapted to dry climates, and would have been available during the ice age.

The Fynbos Biome is flanked on the coastal side by a marine ecosystem with dense and diverse shellfish communities in rocky and sandy intertidal zones (Marean 2010: 427). And as with geophytes, cold seasons do not decrease their numbers. Furthermore, shellfish collecting could have had evolutionary potential since it is typically carried out by women and children, possibly giving the women a new and important role (Marean 2010: 432), not to mention freeing them from relying on men to provide them and their children with high quality food (Marean 2010: 4, Internet copy). Shellfish collection may even have competed with hunting in providing the group with protein.

Large mammals were also scarce in this area, so survival through large mammal hunting was not a viable option. Instead, the predominant species available for hunting there were small antelope, hyrax, dune mole rats, and tortoises. So hunting might have become less important in this environment, even to the point where it became possible to share hunting zones instead of competing over them.

All in all, this area would have been an optimal refuge zone in an ice age. The two ecosystems provided both carbohydrates and proteins which would have been vital for the survival of a human population, be it *ergaster*, *sapiens* or some transitional form.

Shellfish may well have been vital in this transitional process, providing the new humans with important proteins and omega-3 fatty acid. It is even likely that a shift toward more intensive use of marine resources played a causal role in the evolution and spread of modern humans. Omega-3 is difficult to procure from terrestrial resources but is found in abundance in coastal resources. Omega-3 fatty acid fosters the growth of brain tissues prior to birth and in the early years of growth, supporting high rates of brain growth, which may have given major evolutionary advantages to societies consuming coastal resources (Will, Parkington, Kandel, and Conard 2013: 518–19).

Indeed, living off shellfish seems to have become a common survival strategy for early *Homo sapiens*. Several finds testify to this along the South African coastline. Sites featuring the remains of shellfish and other marine products have been found along the west coast, and include Hoedjiespunt and

Ysterfontein, while on the south coast shellfish remains have been found at Klasies, Die Kelders Cave, Blombos Cave, and the eldest sites at Pinnacle Point mentioned above (Will, et al. 2013: 518).

In many ways, these new excavations supplement the early finds at Pinnacle Point. Hoedjiespunt (HDP1), for instance, features abundant stone artifacts, shellfish, terrestrial fauna, ostrich eggshells, and ochre, and it was in use between 130 and 100 ka (Will et al. 2013: 520). Interestingly, the evidence shows that this location was also inhabited by earlier forms of hominine because a tibia and several teeth were found here belonging to what Will and his colleagues describe as a form of *Homo heidelbergensis* (possibly *ergaster*) from around 300 ka, but it also shows that these hominine did not exploit marine resources. Systematic collection and exploitation of marine resources seem to have emerged with modern humans (2013: 534).

At HDP1, a wide variety of stone tools were also found that was made from different raw materials. The tools were made from four major materials: vein quartz, quartz porphyry, calcrete that the researchers could establish as being of local origin because of the large amount of debris also found here, and silicrete, a high quality raw material. Unlike the other three, the silicate was not of local origin but had been brought in from a distance of 10–30 km or even further afield (Will et al. 2013: 523). These objects must in some way have been brought to the locality.

We can therefore establish that Pinnacle Point offered two important resources in its marine ecosystem and Cape Floral Region, namely shellfish and tubes, which belonged to two different zones that were adjacent but separate, and which together provided a livelihood unlike the harsh conditions in the rest of Africa. In Hoedjiespunt, early humans lived off shellfish near the coast but used high quality stone tools that originated elsewhere, so they also used resources from more than one place.

For the south coast, it has been suggested (Parkington 1976) that early humans relied on seasonal mobility to optimize the return of foods that appeared on a seasonal basis at the coast versus the Cape Folded Belt. To account for the nonlocal silicrete at Hoedjiespunt, it has been suggested that the users of these instruments were highly mobile hunter-gatherer groups that occasionally availed themselves of coastal resources as part of their survival strategy, and that they sometimes moved to this area from further away (Will et al. 2013: 533).

The case of Pinnacle Point enables us to test the hypothesis about seasonal migration. According to Marean, this model would mean that the skeletal isotope-chemistry of people who died at the coast would reveal diets similar to those who died in the interior, because the isotopic signature averages out over many years. However, analysis reveals that the coastal skeletons and the interior skeletons had different isotopic values, suggesting that the former enjoyed a diet rich in marine foods compared to the latter, suggesting that the groups were not migratory (Marean 2010: 432).

Marean suggests a slightly different scenario to account for the possibility that the groups supposedly lived separately in different ecological zones. He believes that initially a rather small population lived mostly at the coast, but that when the population began to increase, some were forced into less desirable habitats, and that this eventually lead to two rather separate marine and terrestrial adaptations, and that is this situation is reflected in the isotope signals (Marean 2010: 433). But we must check this scenario against contemporary ethnographic records, which show that people living at the coast need resources from inland (see next subchapter).

We seem to have found a likely location for the emergence of early modern humans, or maybe even the transitional place where they came into being. They were isolated in a relatively small territory, and their numbers were not very high, all in all a very promising situation for speciation. And if this was indeed the place of our origins, our early ancestors would have had to adapt to a new resource situation and to learn to exploit marine resources and supplement them with their more customary diets based on hunting and gathering.

Perhaps the sea really did save humanity, or even bring modern humans into being. This could help explain human dependence on omega-3 fatty acid, a substance which would have had a great impact on brain development at a time when it was urgently needed. Early humans faced by far their biggest challenge ever—to survive in impossible conditions, becoming modern humans in the process. In sum: this place, or somewhere very similar, featured two essential survival resources which grew abundantly but in separate and adjacent zones, and which were not easily exploitable by a nomadic life form, seems a likely location for the final transition into *Homo sapiens*. This situation would give early humans an incentive to share resources between their zones, thereby instigating an early form of trade.

Since the period when Pinnacle Point was first inhabited by early humans, evidence of the presence of modern humans in South Africa seems to pile up. We have ample evidence of human presence at the Still Bay (SB) dating from 77–72 ka, and from Howiesons Poort (HP) dating from 65-59. At SB there have been recoveries of foliate lithic points, engraved ochres and bones, marine shell beads, formal bone tools, and structured use of space. In HP there have been recoveries of backed geometric lithics, advanced projectile systems, engraved ostrich eggshells, bone tools, and there is also evidence of individual domestic hearths (Henshilwood and Dubreuil 2011: 361).

At both sites, bifacial finds represent the uppermost level of technical skills and methods of manufacture and the materials have been carefully selected for best quality and fine grain. As Henshilwood and Dubreuil (2011: 371) posit, this selection may have been carried out with a view to increasing the exchange value of these tools. Such bifacial points very likely served as multifunctional tools and some were used both as spear points and knives (Henshilwood and Dubreuil 2011: 371), and they were therefore also used in combination with other kinds of material that served as handles.

Ochre was also found in all these places. This was used as a pigment and some even bear signs of deliberate engraving. But interestingly, ochre was also used in compound adhesives together with plant gum to haft or attach stone implements to handles (Henshilwood and Dubreuil 2011: 372).

Space also seems to have been partitioned into separate functions that together made up a shelter. Such shelters now featured particular areas for hearths, for sleeping and for tool-making (Henshilwood and Dubreuil 2011: 373). Beads were also found that show their inhabitants' ability to drill and to use cords, and it is evident that these beads were once colored, suggesting that they were used for personal ornamentation (Henshilwood and Dubreuil 2011: 374).

Therefore, the evolution from *ergaster* to modern human also seems to be a move from cooking to combining things in order to survive—or rather supplementing cooking with a combining strategy. For modern humans, combining seems to have become the new fashion: combining resource zones, combining materials, combining compounds. Combining was thenceforth used in all matters of social interaction and production.

But by now this book has exhausted the archeological explanations and genetic insights into the causes of human evolution, and in some senses we are none the wiser. From genetics it is known that all humans have a common ancestor living somewhere in Africa. From archeology it is known that this could be in South Africa and possibly in the Pinnacle Cave region. From archeology it is also known that they, for some reason, began to decorate their utilities and that they left traces of this behavior in Pinnacle Point Cave as early as 164 ka. It is also known from other finds that humans at some point began to trade between groups. Combining things was the trademark of the new species.

THE ETHNOGRAPHIC LEGACY

Ethnographically, we know the result of what happened at Pinnacle Point and in the surrounding area. So our task is now to use our ethnographic knowledge to work backward to the survival struggles that took place here by using logical inference based on the inherited practices of the descendants from this place. We now need to study the ethnographic record with a view to sorting out the nature of the selective pressures that eventually created modern humans.

Human universals, as discussed in the introduction, must have been instigated around the time of the inhabitation of Pinnacle Point, and it could probably have been in this habitat or somewhere that resembled it greatly in terms of resource composition that these traits developed. That which characterizes the present human population must also have been the key to early human survival in this hostile environment—otherwise these populations would not have become *Homo sapiens*. They could just as well have carried on being *ergaster*.

Recent debates emphasize that despite the huge panoply of theoretical stances in the social sciences, all of these essentially seek to understand the same basic features of the human condition. And, as we saw in the introduction most, if not all theories, revolve around the relationships and differences between groups, notably trade and diversity. We thus have to combine our knowledge from the abovementioned archeological finds with the pan-human conditions deduced from current debates about human societies.

We are fully cognizant of the results of the environmental pressures faced by early humans. But since early *Homo sapiens* clearly learned to cope with these challenges, we must assume that their survival strategies must have become part of the human tool-kit that they brought with them as they left Africa: for example, they were common to the new breed.

We can find the answers to how early humans managed these pressures in the ethnographic record. This will tell us how *sapiens* conquered these challenges. To do this, we must turn to anthropology, which is the only discipline that can account for these strange new practices. Then we need to search the ethnographic material for comparable situations and use this as circumstantial evidence of what is likely to have happened in South Africa near Pinnacle Point.

But we should not be looking at hunter-gather behavior or foraging strategies in isolation, as most archeologists have done who have tried to explain human behavior around Pinnacle Point and adjacent places. Modern human groups never live in isolation. They always have several kinds of relationships with surrounding groups. Instead, we should look at interactions among groups, because all groups are part of a reproductive system and their political units are always smaller than their economic sphere of reproduction (e.g., Friedman 1994). Finally, we should be looking for coastal groups and investigating their survival strategies.

Papua New Guinea immediately springs to mind as a good place to study groups in interaction. It was one of the places modern humans reached relatively swiftly after the exodus from Africa around 60,000 years ago. It was colonized relatively late in the eighteenth century, providing a quite complete ethnographic record of pre-contact native practices. It is renowned for its many trading systems, spread across the region. What is more, along the southern coast of Papua there are groups living in conditions that resemble those around Pinnacle Point. These groups also live in a relatively barren coastal environment with a more fertile hinterland.

Among many other groups living along this coast we find the Motu. Motu villages and populations inhabit 125 kilometers of land along the coast, running on both sides of what is now the capital city of Papua New Guinea, Port Moresby, with the heaviest clustering of habitation around Fairfax Harbor.

We have archeological evidence that the Papuan coast has been inhabited with varying intensity for at least 4,000 years, and we also know that the present Motu population arrived at this location from somewhere else some 800 years ago. We also know that they historically made a living in much the same way as the inhabitants did when the first missionaries and colonizers came to the area in the eighteenth century. And the way they cope with the local environment has been amply documented.

The Port Moresby region is characterized by extremely low rainfall compared to other areas in New Guinea, but rainfall increases from the coast as one moves inland, so that 45 km from the coast the mean rainfall increases from less than 100 cm per annum to almost 350 cm per annum inland. This has resulted in six clearly different environmental zones running inland and parallel to the coast, all within 35 kilometers of each other.

In this ecological system, the majority of the Motu live at the coast, with the Koita and Koiari living further inland (Allen 1977: 421). The Motu were principally fishermen and traders, and the Koita gardeners and hunters, a division of labor that resembled the situation around Pinnacle Point with two adjacent ecological zones featuring different resources.

Fishing plays a larger role in the ethnographic material in regard to the Motu, and one which was clearly important all year round. Conversely, hunting and gathering activities are not well documented. Shellfish gathering on the coast is still carried out by the women, and shell middens are common archeological manifestations on this coast (Allen 1977: 428–9).

It is possible to grow some staple crops in these conditions, mostly yams and bananas. Coconuts do not flourish around Port Moresby except in very few locations. Yams and bananas were seasonal but the yield depended on rainfall, and given good rainfall, garden produce could support the Motu until August. Bad rainfall meant famine (Allen 1977: 428). The Motu's own ecological niche did not support food self-sufficiency so they had to compensate somehow for the lack of local resources.

Motu had to maintain access to the meat, vegetables, and other raw materials available inland, and they seem to have done this through alliances and trade (Allen 1977: 448). The ethnographic evidence clearly points to the fact that valuables were exchanged directly for food or otherwise facilitated exchange in many situations. The Motu also equipped trading expeditions along the coast to the Papuan Gulf where they traded pottery for sago. They were forced to go on these sago expeditions in order to supplement the poor food producing potential of their own region which suffered from low rainfall and infertile soils (Allen 1977: 450). So while the Motu were able to exploit the marine environment, this was insufficient for their needs. Everything suggests that it is very difficult for humans to live off a marine zone alone: they must supplement their diets with other resources. In other words, they were forced to trade themselves out of their environmental predicament (Harding 1994).

This Papuan example would appear to be the end product of the struggle for survival around Pinnacle Cave where the populations used the combined resources of the Cape Floral Region and the adjacent coastal region. So we can gain great insight from making parallels between the Motu and the early humans in South Africa. We now have to search for a mechanism that could produce a Papuan-type situation in and around Pinnacle Point.

The humans living here might have started off by living in a location from where they could exploit both resource zones, perhaps through some minor seasonal migration, thereby using the zones as an integrated system that could supply them with the necessary means of survival. They would have sought out a location from where they could access both zones. This would have been typical *ergaster* behavior. They would need territories that catered to all their needs at once, even if only seasonally. But something must have changed their behavior or modern humans would not exist there.

We find some of the answers in the way early humans used cave PP13P. Early in the ice age (MIS 6) the coastline was located within 5–6 km of the site, a distance that had a major influence on the way the cave was inhabited. Five to six km is within a typical foraging radius for coastal foragers, who could exploit shellfish from the rocky shores and at the same time could exploit terrestrial fauna (there is abundant evidence of this at the site). So, people were still hunting and transporting significant amounts of animals back to PP13P (Marean 2010: 438).

However, as the ice age continued the sea continued to recede, and at one point the coastline was located more than 10 km from the site. Burroughs writes that the sea level around 84 kya fell to 25 m below current levels to some 40 m under at around 76 kya, continuing to fall to around 70 m lower than current levels in around 67 to 61 kya (2005: Kindle Loc. 1278–81, 1296–97).

At this time, shellfish at Pinnacle Point were either completely absent or only present in trace amounts, suggesting that humans had settled elsewhere. Finds at the site suggest that early humans seldom visited PP13P but that when they did they were clearly there in order to hunt and manufacture tools (Marean 2010: 438).

With the receding shoreline, the two resource zones became wider apart and they became more and more difficult to access as one zone. This would have posed new challenges to the small population that was isolated in the area and they would have had to organize themselves in a new way in order to accommodate the distance between their resource bases.

At this point, some of them found themselves in much the same situation as the Motu living by the coast in Papua, and their solution to the problem is likely to have been similar to the Motu's, since the latter would have inherited a common human means of tackling the problem of limited resources.

Those living in South Africa must somehow have invented or at least used the same solutions that the Motu used to solve the problem of living in a place that could not supply resources all year round. For the Motu to survive in their niche, they had to trade with inland and other groups, and as the distance between the marine and Fynbos resource regions widened in South Africa, the groups would have had to split up and been forced to exchange resources between the two zones.

And here we have the basic situation leading to kinship. At this point in history, stable relationships between groups could only be based on personal relations. A possible scenario for this development is that the first arrivals in the Cape Floral Region were living in groups in places that allowed them to exploit both the marine and the Fynbos Biome resources. Perhaps they would split up during the day, sending some parties to collect shellfish along the coast and others to collect tubers inland, only reuniting during the evening to eat and sleep together. As the coastline receded and the distance between the coast and the Fynbos Biome became larger, this scenario may eventually have changed so that they met intermittingly during the day but slept apart, though they would still have constituted a group with family ties between them.

This could have gone on for a while as the distance grew until they eventually became two separate groups living apart and meeting occasionally to exchange resources from their respective zones, still recognizing their family ties. Instead of becoming two separate groups each foraging in their separate niches, they were still dependent on each other, so they resembled each other and worked together, but lived in different ways. Over time, this produced shellfish-people and tuber-people that depended on each other to survive. Their initial place of habitation would have changed from being a sleeping place to a meeting place, and eventually the meeting place would have become a trading place. In this way, early humans would have been forced to share resources between groups in order to survive. They survived in harsh conditions by inventing kinship—a new way of exploiting resources that had previously been geographically separated. Natural selection would have favored those who were good at sharing and keeping up ties between adjacent groups; the rest would have starved and become extinct.

This situation would be of immense importance for the evolution of these early humans into *Homo sapiens*. The receding seashore gave an incentive both to share and combine: both acts are the first of their kind in the animal kingdom. Humans would share their hunting grounds and they would combine resources from two different environments, which, in combination, would secure them a livelihood in harsh times. At the same time, they would be exchanging one product for another. This could be seen as a proto-symbolic act.

As soon as a group starts to exchange or trade with other groups, the rules of the game change. Then groups are no longer confined to their own resources. They have become part of a network, and from then on it is the network that determines the conditions and characteristics of the singular groups. They no longer forage on their own in their own confined territories, and the necessary resources for survival no longer have to be available within their own range.

It is here that we find the great evolutionary leap forward. Humans changed their behavior in response to a changing environment. Pure adaptation. They finally broke out of their dependency on the constraints of the immediate environment. To be human now meant to be able to connect, or combine, ecological niches. Once this principle had been invented, it could be extended endlessly to all kinds of phenomena and items: to tools and implements, to substances and materials, producing ever more sophisticated objects and social arrangements. To be human meant to be able to combine things and make alliances. As a result, sheer strength gave way to tactical maneuvering. Humans became graceful, slender, and smart.

So the receding seashore eventually gave the population living there an incentive to exchange. This had become a matter of survival. If this scenario is correct, the seashore served as an evolutionary elastic band, drawing resources incrementally and perhaps imperceptibly further apart until groups had to split into subgroups that eventually came to live apart. In this way the sea both saved and changed humanity. This happened slowly and incrementally over many decades during which humans' ability to plan steadily increased as they coordinated meetings and other exchange situations. Natural selection would have benefited those who were good at making alliances, and it was they who would have survived in this new situation.

In this new situation, humans came to live in more permanent groups in each ecological zone, where they supplemented their diets with resources from the other zones: shellfish people–tuber people. These humans belonged to the same group but lived differently and apart. And perhaps it became important for them to signify where they belonged and the differences between them. This is probably what they used ochre for. These were the first steps toward the emergence of what are today called ethnic groups.

Ethnic groups are a much debated subject within anthropology, but there are some particularly relevant studies that may help to shed light on the sudden appearance of ochre at places such as Pinnacle Point, and its subsequent continued use by early humans. Ethnic groups must have originated somehow, and the situation in South Africa was perfectly suited to producing this kind of social organization. Ethnic groups strive to signal differences despite their similarities. They seek to express belonging and identity.

We can usefully draw on Frederik Barth's conception of ethnic groups and boundaries (1969) in thinking about this issue. Barth maintains that cultural differences are not intrinsic to groups living apart and in isolation, who come in contact with one another and wonder how the other group became so strange. For Barth, ethnic distinctions do not depend on the absence of social interactions and acceptance (1969: 10). On the contrary, these are very often the foundation on which social systems are built (1969: 10). Ethnic groups are categories of ascription and identification by the actors themselves, and they therefore serve to organize interaction between people. But as we shall see, this is not the whole story about ethnic and cultural differences.

From this it follows that a group that splits in two to exploit adjacent resource bases is forced to trade its zone-specific products for the other group's, and as a result comes into contact with them, and has an incentive to ascribe characteristics both to its own group and to its trading partner. This would be a first step in organizing interaction between two such groups. Not only that, it would place them firmly on the road to culture.

As Barth points out, such groups are maintained despite changing participation and membership in the course of individual life histories (1969: 10). It is the *borders* between them that are important, not the group membership per se. This idea is very fitting for a group, such as the early humans in question, that was slowly splitting apart but which kept changing its membership since it was originally based on foraging parties.

It is also clear that since these groups occupy different niches, one cannot ignore the problems of numbers and balance in their adaptation. Whenever a population is dependent on exploiting a niche in nature, there is a natural upper limit on the size of the group the niche can sustain. When the group size increases, the relationship between the groups changes, and in order to remain within carrying distance, part of the group might have had to migrate so it could function as an outpost settlement from which it could link to the old group. Alternatively, it could perhaps try to establish a network in the new locality. As the population grew and times improved after the ice age, the group could then start to leap-frog into other niches and thereby colonize large areas.

In principle, the position of such groups in their network could have been based on any kind of resource that was in demand by other groups, for instance a particular foodstuff, a specialized product like high quality tools and instruments, or vessels or valuables that other groups found attractive. In this way, early humans would have been able to increase the combination of resources at their disposal in ever expanding networks of exchange.

To be able to live in networks and trade between groups, early humans would also have needed to develop networked brains. And indeed, this was a time when the human frontal lobe grew. This lobe serves as a kind of executive officer that makes decisions, sets goals, plans, organizes, and formulates strategies, and it is connected to the ability to integrate ideas across space and time, and to sequential memory function (Coolidge and Wynn 2009: 553).

By now we have charted the full sequence of human evolution and the different skills that characterized each period. First, humans evolved bipedality when they came down from the trees and started to live on the savannah. Next, they became more mobile and learned to run and make fire so they could live on the open savannah. Finally, they learned to combine ecological zones by trading with each other. Each of these abilities developed in response to the changing climate and human adaptation to a new situation that subsequently widened the ecological niche of the species.

Human evolution therefore presumably had nothing to do with making tools, or with the mental activity purported to be necessary for this. Tools were merely derivative effects of specific adaptations, and they served a limited purpose for very long periods of time. Neither did human evolution have anything to do with steadily increasing sociality or excessive grooming. These practices were merely adaptations by earlier forms of the human species. If the groups became too big they would, logically, have split up into identical but smaller groups.

But both tools and sociability became stepping stones that the next species could build on and modify to fit with new adaptations. Having been forced to live in separate groups that were initially one, these groups still had family ties between them. To be able to continue to exchange, they would have had to renew family ties in each new generation. They were well on the way to inventing kinship, the most universal feature of *Homo sapiens*. Kinship became the organizing principle behind the new ability to combine niches and trade one group's products for those of another, because kinship incorporates both differences and alliances.

Chapter Three

Kinship and Exchange

The last ice age dried up much of the African savannah and blocked the exit routes out of Africa, leaving behind only small pockets of habitable areas with isolated human populations. All this produced extremely favorable conditions for speciation. One of these pockets was located in South Africa in the Cape Floral Region that contained the so-called Fynbos Biome, characterized by an abundance of tubers, and flanked by a seashore with plenty of marine resources, including shellfish. This place may well have been the birthplace of modern humans. If not, it must at least have posed the same survival challenges as those faced by the first humans.

To survive, this group invented kinship to create lasting relationships between groups who lived in adjacent but different ecological niches, so that they could exchange products with kin. This was probably a purely adaptive response.

It is from this area in South Africa that we have the earliest finds of modern humans, who seem to have arrived here during the last ice age. In this area that was relatively untouched by the surrounding mega drought, the (emerging) human population could utilize resources both from the sea and from the Fynbos Biome. However, the seashore kept receding, drawing the respective resource bases farther apart and posing a new challenge for the human population by slowly separating them into groups living in separate zones.

But given that it would have been almost impossible to live off marine resources alone, as we know from Papua New Guinea, the coastal groups were forced to find another way to keep utilizing both resource bases. So, instead of living together in one place from where they could send foraging parties out in each direction, they had to find a way to cope with the fact that the distance between the two resource niches became wider and wider, possibly leading to a situation where they slept apart but met to share resources.

In this new environment, the main available resources were shellfish and tubers, both of which were collected by females, while hunting became a more precarious task and, as the finds from cave PP13P indicate, game mainly consisted of small animals. Females therefore became more important for the group's survival, and they may even have become the group's main asset. Despite the fact that the groups had to maintain ties in order to continue exchanging resources, it must also have become less advantageous for them to send their valuable females away to mate in other groups.

So this ice age ushered in completely new selective pressures on early human groups. The practice of sharing resources between groups meant that early humans began to live in networks that enabled them to transcend the limitations of their immediate environment. By this stage, they were exchanging resources between ecological niches.

Initially, exchange was probably just a way of sharing resources between connected kin groups. As the distance widened between groups, and more groups got involved, such exchanges probably became a form of barter. It was not as if they started out as businessmen trading stocks and shares in suits and ties. But it was crucial for them to maintain steady relationships between groups in order to secure their continued survival. The groups in question needed some form of stable and reliable connection that connected them according to some already latent principle. The anthropological records teach us that the most widespread form of human social organization is kinship, so we may assume that this was how the early humans organized the relationships between their groups.

Probably one of the best places to study kinship as a form of inter-group social organization is among the Australian Aborigines. The Aborigines are perhaps the most direct descendants of the early modern humans who emerged from Africa, and the ice age in Africa must have produced living conditions that resembled those that the Aboriginals had to contend with almost all year round, especially in the desert areas, and indeed they seem already to have been well-adapted to cope with this situation.

Aboriginal society has been intensively studied. Scholars have focused particularly on their different descent systems. Aboriginal society had systems in which each person belonged to a certain subsection of a clan that determined what subsection their children would belong to and who they could marry. There have been intense debates about how to understand the seemingly very complicated kinship system that the Aborigines invented, notably between Radcliffe-Brown (1931), Elkin (1940), and Lévi-Strauss (1969), who all disagreed about the interplay of marriage rules and descent. This debate centered mainly on the formal aspects of kinship as an abstract system and not its adaptive role in the lives of the Aborigines.

Yet we know that many Aboriginal societies lived in very harsh environments, and that the Aboriginal economy both in the interior desert areas and in the tropical coastal areas was based on the gathering of vegetable food, which composed 70 to 80 percent of their diet. Under these circumstances, meat becomes an intermittent phenomenon, and a larger hunting zone is necessary (Yengoyan 1979: 397).

Aborigine clan systems were also a means of coping with the environment. Since drought is a recurrent phenomenon in Australia, groups had to maintain ties with other groups that could be activated during periods of stress (Yengoyan 1979: 399). This has been called a "production group diversity system," in which different production groups are created by combining patrilinearity and marriage, linking together the different groups for the purpose of resource exploitation (Yengoyan 1979: 401). In this way, each group had a core niche that it could exploit, and it could control other groups' access to that niche. Each group also had access to a larger common "hinterland" which it shared with other groups. Given the limited number of game animals, core zones seem to have been the Aborigines' main gathering area, while the common area extended the zone any one group could utilize, resulting in a larger hunting area.

We know that this was the end result of early human evolution, but we still need to envisage a scenario that can account for it. As a result of the diminishing number of game animals, gathering must have become humans' main means of obtaining food, supplemented by the occasional game animal. Just like in Australia, 70 to 80 percent of nutrition was probably procured by female gatherers, be it tubers or shellfish.

In a situation where gathering becomes the main means of nutrition as, for instance, in the Cape Floral Region, females become an asset. When times are harsh, this can lead to a seemingly contradictory group survival strategy: do your best to prevent young females from leaving the group, and make sure that your young males find a female partner fast. This is not feasible in the long run. But it would be a great incentive for a group that had lost its daughters (especially if it didn't have many in the first place) to demand something in return for its lost workforce. Given that there wasn't much game around, you would probably want access to a larger hunting zone or the products of the females in the other group, especially if these could supplement your own resources.

Having already inherited a notion of kinship, the logic of survival in a situation of diminishing resources, or resource separation, is simple: share your resources with your kin. There was already some kind of division of labor between the sexes, namely male hunting and female gathering, so all that was needed now was to extend this division of labor to the group level. So, to escape their survival predicament, the new *Homo sapiens* invented an incipient form of marriage (i.e., exchange of partners), turning single-group foraging into alliance networks. This is, arguably, the only way we can account for the Aboriginal kinship system. Those who excelled at establishing alliances had a better chance of survival in these harsh conditions.

Instead of foraging as singular groups *within* their immediate environment, they invented exchange relations *between* groups based on alliances created through wife exchange. In a resource-segregated environment, these groups became embedded in *networks of kinship* which foraged in an interconnected territory. Relying on such networks also meant that it became possible to live by the coast and utilize seafood resources on a long-term basis, since coastal groups could procure other needed products (obtained through gathering) through exchange with their kin further inland.

In this way, these groups could live in territories that did not provide all the resources they needed for survival, as long as they could utilize adjacent territories by maintaining relationships with other nearby groups that had access to supplementary resources. In other words: they changed their behavior to adapt to the new environment. Merely by maintaining close bonds with their daughters (even if they lived in another group), they could activate an already existing, though dormant, matrilineal connection between groups through which resources and obligations could travel. They had discovered how to exchange resources through personal ties, transforming them from autonomous groups responsible for their own survival into groups that shared and exchanged resources in alliance-based networks.

This also meant that they would have developed a special relationship to their daughters' children. Thus, in one fell swoop, these groups invented descent and marriage as a means to establish lasting reciprocal bonds with other groups. Living in alliance networks "solidified" relationships between groups and between the sexes, making them lasting partners. Males were no longer the protectors of the group, but rather negotiators *vis à vis* other groups. So, by combining descent and alliance, they were able to extend their biological niche.

In the process, they had to develop new ways of conceptualizing their environment to be able to envisage and track their alliance partners' shifting locations. Forging alliances through kinship relationships, and keeping track of these, obviously favored those that had the incipient mental capabilities to see the larger picture in "map-like" fashion. They also had to learn how to interact within a networked environment. And they needed to learn to control their emotions toward people from other groups. These are all cognitive operations that are associated with the frontal lobe of the cerebral cortex—which is known to have grown at this time. This changed *Homo sapiens* morphologically: we were now born to think.

EXCHANGE AND SOCIAL RELATIONS

Having instigated kinship and exchange between groups, early humans stepped out of their immediate environment and came to live in networks of reproduction. What they could not find in their own niche they could obtain through trade. In this way, relationships between groups gained priority over direct foraging. This brought an entirely new dynamic to the human population. As the population increased, it could connect more and more niches based on groups that were interlinked but whom—for that very reason—emphasized their differences and expressed them symbolically.

But the ethnographic record also reveals that the modern descendants of these groups are organized socially in differently constituted groups, giving the impression that they must have evolved their social organizations in isolation from one another. This has given rise to debates about the primordial organization, about the origins of kinship systems, about whether social organization was based on matrilinearity or patrilinearity, and about the mechanisms through which one or more of these changed to produce today's immense cultural variation.

Barnard, for instance, thinks that there were potentially two types of kinship system that could have been the primordial ones. One is mainly found in Africa and is characterized by flexibility in social organization, in kinship, and in the symbolic domain. The other is mainly found in Australia and is highly structured, featuring very strict marriage rules. So, primal society must have been either fundamentally flexible or highly structured. Barnard believes that one cannot have it both ways (Barnard 2012: 53–58).

There are some major differences between matrilineal and patrilineal societies that should be taken into account in relation to early human development. In matrilineal societies, the bond between brother and sister is strong while that between husband and wife is weak; children belong to their mother's kin group and children inherit from their mother's kin group, unlike in patrilineal groups.

Generally, matrilinearity and patrilinearity are seen as ideological reflections of more objective conditions, so that societies where cooperation among men in production or military activities is important tend to be patrilinear, while societies in which cooperation among women is more important will tend to be matrilinear. Steward (1936) argued that among early humans, men's hunting was all important to the survival of the groups and therefore sons inherited their fathers' hunting grounds, which probably made them patrilinear. Opie and Power (2011) argue that the role of grandmothers as foragers and caregivers to their daughter's children was critical and therefore that the groups probably coalesced around matrilineages.

But these claims are not based on an in-depth understanding of kinship. They assume that the groups were foraging independently and almost in isolation and that they somehow grew more and more sociable until they developed kinship. Kinship in this rendition would not serve any purpose, except showing that early humans acquired it. There is no adaptive reason for kinship to evolve in this situation. It is just supposed that the human lineage at some point began to evolve on auto-pilot, propeling human evolution forward because of some inbuilt sociability. But adaptation does not work that way.

At the same time, we know that kinship is the most fundamental human institution. So kinship must have been vital to the transition into modern human-kind. Still, kinship is only necessary if you want to keep track of your origins when you change groups. Descent and alliance are two sides of the kinship coin. Everything is about where you come from (or from whom) and where (in what group) you live, and this information is only necessary if there is an ongoing relationship between the groups. So kinship is about relations between groups. Without kinship, all we have are parallel groups foraging independently in their respective niches.

This also shows the importance of what has been termed "the elementary unit of kinship," which does not, as is often supposed, refer to the nuclear family, but the nuclear family plus the mother's brother, since the nuclear family can only reproduce itself through the exchange of women and so the most minimal kinship unit must include the mother's brother (MB), the person who gave our group a wife and means of procreation. As Lévi-Straus writes: "I will give up my daughter or my sister only on condition that my neighbor does the same. . . but the fact that I can obtain a wife is, in the final analysis, the consequence of the fact that a brother or father has given her up" (1969: 62). This is the basic "atom of kinship": man—wife—the uncle who gave the wife. Not until this happens do we have kinship.

Given the very small initial human population, we must assume that they must all have started out with the same basic equipment and a very similar form of social organization. The answer is to be found in the abovementioned "atom of kinship." Accordingly, differences between groups and their social organizations must relate somehow to the basic atom of kinship, and all these different forms of social organization must derive from "the atom of kinship." It should, therefore, be possible to deduce different kinship systems from this.

It may seem improbable that the elementary unit of kinship can account for all the diversity we find in social organization and cultural differences. But this is actually the case. In a very basic way, the elementary unit of kinship is the algorithm upon which all human organization is built.

As I have learned from my professor, Jonathan Friedman, and which the following is based upon, kinship is basically based on exogamy and the exchange of women, and this situation divides the primary sibling group (brothers and sisters) into two groups: those that "leave" and those that "stay." So the rule of exogamy requires that males and females pursue different marital careers. As a logical extension of this, exogamy also divides cousins into what are known as parallel and cross cousins, i.e., those children who are related through siblings of the same sex as father's brother's children (FBch), and those related through mother's sister's children (MZch); and on the other hand, those related through siblings of the opposite sex through father's sister's children (FZch) and those related though mother's brother's children (MBch), and this applies to second, third, and all other cousins. This means that parallel cousins belong to the same unilinear grouping as siblings, while cross cousins belong to a separate unilinear grouping, i.e., the one into which one can marry. Schematically, it looks like this:

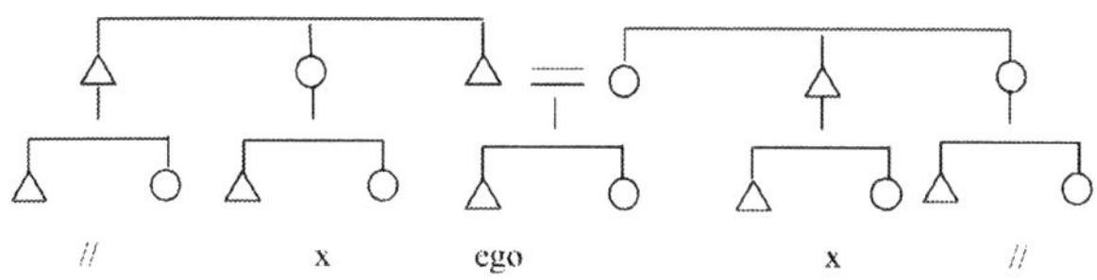

△ male; ○ female; = marriage; siblings; // parallel cousins; x cross cousins

This illustrates that cross cousins live in other groups. This example shows that if you are a female offspring, your father and his brother live in the same group as you, while his sister will have moved to live in another group. Parallel cousins also live in that group, while your cross cousins live in your father's sister's group.

So, the elementary structure of kinship determines that parallel and cross cousins live in separate locations, and serve different purposes in the overall kinship structure. This is the elementary relationship between groups: for example, two groups that exchange spouses. This is the elementary structure of human social organization, which may be depicted as follows:

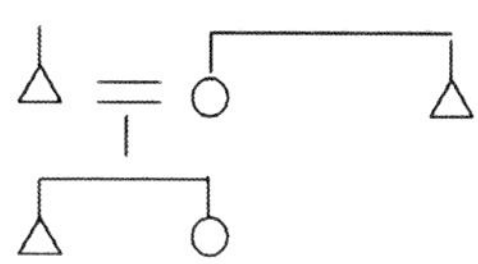

Nuclear family + Mother's Brother = elementary unit

△ male; ○ female; = marriage; siblings

This was all early humans needed to invent, and all other (tribal) forms of social organization follow from this. This is the basic structure of kinship, but seen through an anthropological lens it is not kinship or marriage that determines the relations between groups. On the contrary, it is exchange or alliance relations that determine the distribution of kinship categories, which is just to say that kinship is not endogenous to groups but a result of the character of the exchange relations between groups. So the kinship system in any one group is not an inherent feature of a local culture that can be studied on its own. It is all about relationships. This is demonstrated by the fact that kinship category terms such as mother's brother's daughter (MBd) are applied not just to real cousins, but to all appropriate members of a group that relate to my group as "wife-givers."

As a direct result of the different marital careers of sisters and brothers and the implications this has for the distribution of cousins between groups, it is intergroup relations that determine what kinds of cousins one can marry. Therefore, the term "cousin" translates directly into exchange relations. We find three forms of cross-cousin marriages: bilateral, matrilateral, and patrilateral cross-cousins that correspond to three elementary structures of exchange, namely: restricted exchange, generalized matrilateral exchange, and generalized patrilateral exchange, in which patrilateral exchange produces short cycles of exchange while matrilateral exchange produces long cycles of exchange.

Although it may seem that a rigorous anthropological training is required to understand all this, it is actually quite simple. All one has to do is focus on the relationship between groups. In bilateral cross-cousin marriage, father and mother's brother live in two different groups and they simply exchange sisters. This relationship is based on reciprocity between two groups, and therefore the exchange system is restricted to these two groups.

In matrilateral cross-cousin marriage you marry your mother's brother's daughter (MBd) and she comes from the same group as the group your mother came from, for example, my MB's group. This does not involve reciprocity but a one-way exchange of women between groups, so that the same group will always be wife-givers to the same recipient group. This implies long cycles of exchange with at least three groups, so that your MB and MBs can also find wives, or else problems will arise at both ends of the exchange system (in fact, there are many empirical examples of this arising, to which we will return shortly):

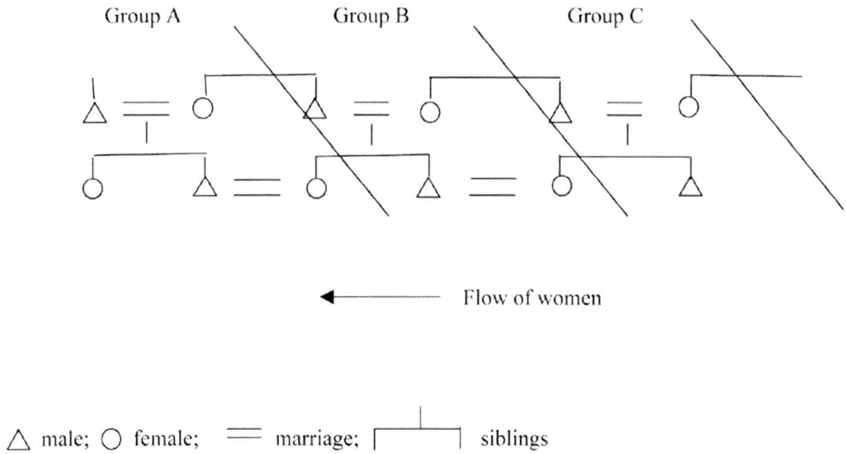

Or with kinship terms:

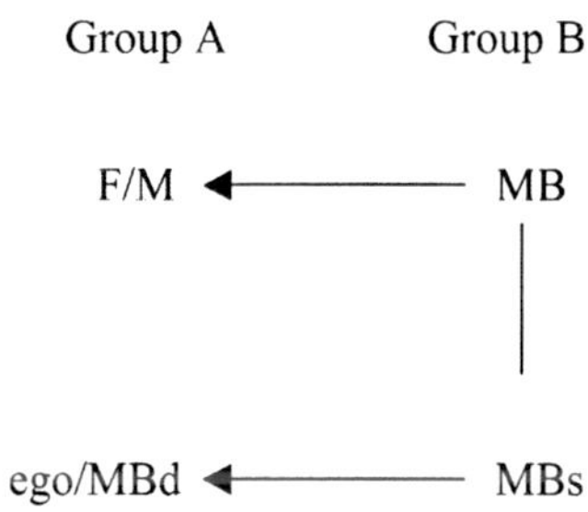

In patrilateral cross-cousin marriage you marry your father's sister's daughter (MZd). When you marry your FZd it means that you marry a woman from the group into which your FZ married and this implies taking a woman from a group to which your group gave a woman in the previous generation. Consequently, the flow of women is reversed in each generation.

This prodes the following relationships between groups:

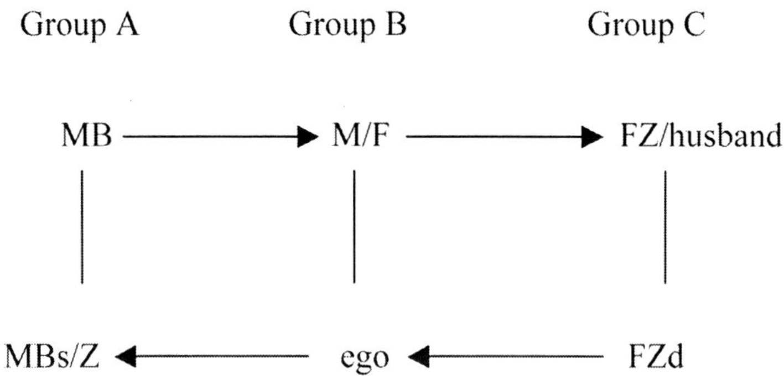

This shows that the basic element in kinship is the relationship between groups, which is expressed in kinship terms. If we suppose that one of the abovementioned kinship systems is the primordial one (and one of them has to be), it must be the one with two groups that exchange sisters in what we have term "restricted exchange" based on reciprocity between two groups. This could account for a situation in which a group is slowly pulled apart, eventually becoming two separate groups that exchange wives (as could well have happened in the Cape Floral Region). We would then have a form of sister-exchange and restricted exchange based on reciprocity and a more direct exchange of resources.

All these kinship systems can, thus, be seen as different ways of negotiating the management of resources. Kinship systems also allow for specialist groups to be incorporated in their exchange networks, thus enabling the incorporation of different food regimes, and also making it possible to supplement these regimes through trade with groups producing special products like fine or specialized tools and who, therefore, did not have to rely on their own food resources. This was a brave new world.

However complicated or controversial different kinship systems appear to be, and irrespective of whether they are Australian or African, they are all, as Friedman says, variations of these three systems. For instance, the seemingly very complicated Australian kinship systems that have confounded anthropologists for centuries are all variations of restricted exchange (see, e.g., Allen 2000: 82–3). For instance, among the Kariera tribe there are four sections:

> Named Banaka, Burung, Karimera and Palyeri. A man of one section may only marry a woman of one other particular section. Thus a Banaka man may only marry a Burung woman. The children belong to a section different from that of either the father or the mother. The children of a Banaka man and a Burung woman are Palyeri, and they in their turn may only marry with Karimera. (Radcliffe-Brown and Kuper 1977: 134–5)

This seems very complicated, but the complexity of Australian kinship systems is largely an illusion produced by the existence of large numbers of named groups and kinship terms.

However, the underlying principles that generate that complexity are extremely simple—otherwise the early Australians could not have managed it. For the Kariera, the four-section system results from the reciprocal exchange of women and the division of the groups into two generations. Basically, it is all about reciprocal sister exchange, but between four sections, and therefore the sister exchange is reversed between the groups in every generation. So they give the sections names to remember the order of exchange.

These considerations clearly show that kinship is the primordial means of organizing exchange between groups. Now that we have established this, it becomes clear that the question of whether these early groups were matrilineal or patrilineal must also relate to exchange. In evolutionary terms, the difference in linearity has been discussed in relation to survival strategies for single groups, but with the advent of kinship, survival became a networked strategy, and thereby linearity must also be, or have quickly become, a derivative of intergroup exchange systems and their overall organization.

We know that at that point in human evolution, women became immensely important in securing the survival of their groups, but their role (probably) did not produce matrilineal societies. Rather, women's value as anchors of the groups' survival was a key factor in producing kinship, the most universal of all human institutions. Kinship and alliances have long been treated as two separate features of human organization, but really they are just two sides of the same coin when it comes to living in a networked mode of reproduction. Consequently, linearity must be linked to this form of alliance network.

Material from West Africa suggests that the categories of matrilinearity and patrilinearity are not quite as significant as they have been made out to be (Leach 1961), and this is especially clear from the ethnography of the Kongo Kingdom located in Central Africa (this kingdom broke down soon after European contact in the mid-seventeenth century) (Friedman 2008: 233–34). Congo-type societies like the Kongo Kingdom from this period have been regarded as matrilineal societies because the micro-unit consisted of so many matrilineal descent groups and therefore it was assumed that the macro group had the same structure as its constituent parts (Friedman 2008: 236).

But as we now know, kinship is about networks and therefore it is intergroup relations that are essential in establishing local conditions. For that reason, we have to reverse this causality between micro- and macro- group levels and look at how the macro group as such consists both of specific subgroups and a structure which unites them, giving an internal dynamic to the system as a whole. In this macro system, we have kinship-based subunits and a political structure that unites them, and the latter is intimately related to the possibility of accumulating the prestige goods necessary for the social reproduction of individuals, since these are used for bride price and also enable the elders to establish and maintain dominance over the younger members of society (Friedman 2008: 231–36).

In the Kongo Kingdom system, groups sent out their sons to be married in subclans to which they paid bride price, and who in turn sent their sons off to be married in another subclan where they paid bride price, and in every instance this was reciprocated through some kind of tribute. This was made possible by a kinship system that was based on the elementary unit of kinship, so that the sons were sent out married their mother's brother's daughters (MBd). This meant that the same group was in a constant position as wife-giver to the same recipient group across generations. It also meant that women, so to speak, married upward in the system and were accompanied by tribute, while prestige goods traveled in the other direction, giving tribute and redistribution of equivalent but different products (Friedman 2008: 244). In schematic form, the Congo system looks like this:

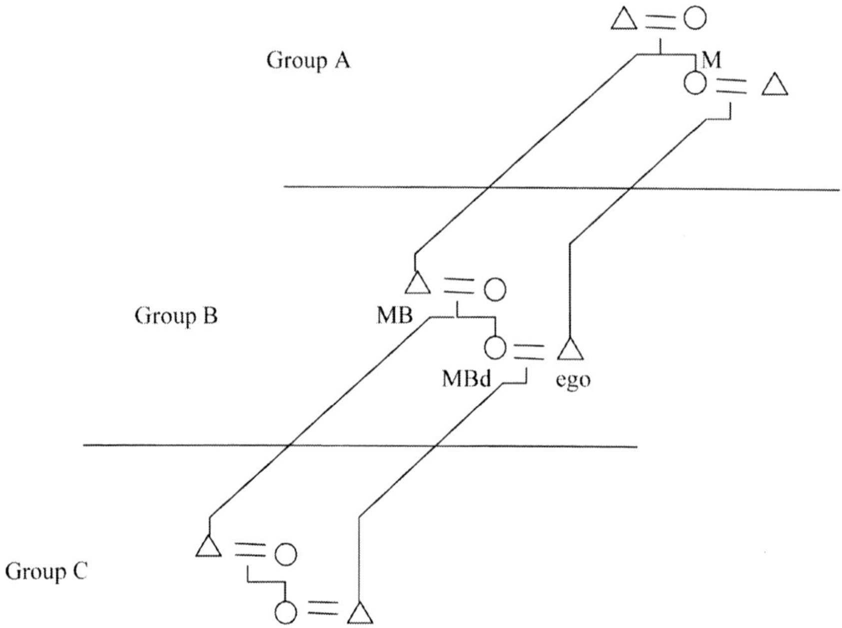

This is a more sophisticated version of matrilateral cross-cousin marriage but with a special residential component, because the man moves from his natal group to live in his uncle's group. The terminology for this kinship system is *matrilateral cross-cousin marriage with avuncular residence*. In this Kongo case, we see that the appearance of local matrilineal groups was the result of a political structure in which men moved out, producing regional patrilines that also produced a political power structure that integrated the whole system. As Ekholm Friedman writes, "the localized segment of the matrilineage . . . has, in fact, no independent structural existence" (2008: 236). And consequently the question of matrilinearity versus patrilinearity is dependent on how the elementary structure of kinship relates to local resource exploitation, and on the possible presence of prestige goods in intergroup relations.

So it is now clear that kinship is what made us human. Kinship did not arise due to a steady growth in sociability, but as an adaptive mechanism that enabled humans to mediate between ecological zones. To ensure stable relations, they needed to know from which groups their members came, otherwise lasting alliances would not be possible, and life would have been difficult under the very harsh conditions during the last ice age. Basically, they would have had to stick to *ergaster*-type foraging techniques, which would most probably have led to their extinction.

Kinship is the most basic human condition. Kinship is about managing resources, and about distributing resources between groups. By inventing kinship, early humans came to live in networks of reproduction. This means that the management of resources, so to speak, produced the organization and distribution of local groups, and it is the network that gave these local groups their characteristics.

This is the primordial social situation for early humans, the basic social condition that incorporates exchange and "cultural differences." After the invention of kinship, groups no longer reproduced themselves as isolated groups. This is the primordial condition of humanity, and it must have been the form of organization that made it possible for the human population to expand to every corner of the Earth.

Chapter Four

From Kinship to Culture

Some may think it improbable that kinship is at the root of culture. Others might even claim that it must have been the other way around. After all, how could such a simple mechanism produce all the splendid cultural variations we see today, with all their marvelous symbolic constructions, and all so-called religious elements such as ceremonies, myths, magic, and beliefs in gods and devils?

But it is precisely the way in which culture and religion have been conceptualized that prevents us from realizing the role played by kinship in all this. Conceptualizing cultures in holistic and relativistic terms makes the transition from *ergaster* into full-fledged symbolic life seem almost impossible—almost as though it were due to supernatural means or a self-propelled process unique to humans. And even though we are living proof that *something* extraordinary happened, humans are probably not all that exceptional in their evolutionary history. Rather, it may be the way that culture and religion have been conceptualized that is erroneous.

To start with, culture must have an elementary form—some kind of algorithm that can produce endless variations. From the anthropological records, we know that the most common feature of human social organization is kinship, and that this, therefore, must have played a vital role in the adaptive changes that the remaining *ergaster* population went through during the last major ice age.

Prior to *Homo sapiens*, all groups were organized in separate and autonomous groups who had to find all their resources within their local territories. With the help of kinship, *Homo sapiens* was able transcend this situation. Kinship is all about alliances between groups. Making families was an adaptive mechanism to survive in new climatic and resource conditions. Kinship made it possible to share resources between environmental niches when those resources suddenly disappeared from one's own locality. This genius invention liberated *Homo sapiens* from the restrictions of the immediate environment.

It is commonly believed that at this point in human history, biological evolution ceased to influence the human species, and cultural evolution took over (Barnard 2011: 104). With this, everything became cultural, and the archeological finds seem to corroborate this idea. Suddenly, humans used ochre for decoration and beads for adornment, and they started to bury their dead together with gifts buried alongside them. This is often regarded as an indication that they believed in an after-life; for example, people did not just die, they had become culturally significant and were venerated by the living.

So for some, not only did they acquire culture, but burying the dead seems to indicate that they also became religious. Some of the earliest known burial sites are 95,000 years old, and Wilson thinks that at this time people must have begun to ask themselves where all these dead people went. For him, the answer is clear, they thought the departed still lived somewhere because they saw them in their dreams, where they apparently lived with all kinds of imaginary creatures (Wilson 2012: 264). Therefore, Wilson claims, humans had by now come to believe in a supernatural world.

Culture, then, seems to have coevolved with religion, or at least with something religious, like a belief in an afterlife and the supernatural. It seems that humans' newly acquired ability for symbolism was immediately used to single out a specific area of social life, that was then imbued with extraordinary symbolism. Religion is obviously a subgroup of culture, but we need to understand both phenomena to account for a possible coevolution. However, neither of these phenomena are straightforward. In the social sciences, a prolonged debate has raged about the nature and role both of culture and religion in human life.

TOWARD ELEMENTARY CULTURE

Traditionally, culture has been defined along the following lines:

> The total complex of spiritual, material, intellectual and emotional features that characterize a society or social group. Culture is mental structures, the general patterns of understanding prevalent in a given social group: conceptions of the world, religion, gender role patterns and the management of natural resources . . . culture as the sum of social practices in the form of, for example, religion, language, education and social and family practices. (Danida 2009: 5)

But culture and religion cannot have evolved in this or any other holistic form, so we must try to simplify these important concepts. Classical theories about culture and religion are based on two basic assumptions: both phenomena are integrative mechanisms, and they result in ordered systems of categories. Historically, culture has been depicted as a simple mechanical system consisting of units, where there was thought to be some form of correspondence between the cultural order and individual practices. In this way, "culture" can seemingly be reified into integrated units with inherent traits, such as political organization, religion, and gender roles.

At the same time, this widespread and popular conception of culture has increasingly been questioned by anthropologists. In this traditional form culture has, as Barth observes, become an all-encompassing description of human behavior. He thinks this leads to some perplexing paradoxes when one tries to make generalizations or theoretical statements about the "nature of culture," which often end up seeming contradictory (2002: 24), and which make it almost impossible to fulfil the cultural promise of understanding and regulating human organization. He gives a list of examples:

- Culture is received tradition, but it is also emergent and in flux.
- Culture is shared but also distributed among members of society.
- Culture is associated with distinct societies, but it also shows continuous variations.
- Culture is a complex whole, but it also consists of shreds and patches.
- Lastly, culture is a depiction of a life-way, but also a directive force on human action. (Barth 2002: 24–5)

If not directly confusing, these statements about culture at least show that the concept is indeed an aggregate construction, "covering and combining many diverse phenomena" (Barth 2002: 26), which in themselves make it completely impossible to say anything meaningful about how culture came into being. In this way, Barth points to the difficulties in traditional ways of understanding culture, which has come to suggest boundedness, coherence and homogeneity, features that many critics of the concept find to be in stark contrast to the observable facts of actual practiced life, which is characterized by inconsistencies, conflicts and contradictions.

So, these days "culture" in its traditional understanding is increasingly being held to task for misrepresenting people and their ways of life instead of explaining it. Culture in this guise is now found to be a substance that, in the words of Clifford:

> Orders phenomena in ways that privilege the coherent, balanced, and "authentic" aspects of shared life. . . . Culture is enduring, traditional, structural (rather than contingent, syncretic, historical). Culture is a process of ordering, not of disruption. (Clifford 1988: 232, 235)

As a logical consequence of all these difficulties with "culture," he then finds that "it may be true that the culture concept has served its time" (1988: 274).

As we can see, the classical view of culture has been accused of totalizing people's actions and concealing differences by making it look as though people follow cultural codes almost blindly. This makes it hard to account for cultural change, cultural diversity, and the heterogeneous nature of cultural orders. This view of culture also makes it extremely difficult to find an evolutionary mechanism that brought it into being. It is also incompatible with the abovementioned networked character of group relationships that accompanied the establishment of kinship as an adaptation that made it possible to connect environmental niches, thereby setting groups free from the constraints of the immediate environment. Human groups must initially have been interconnected and have influenced each other's "cultural" constitution.

Appadurai (1996) alludes to some of the more basic characteristics of culture. He questions whether any shared culture exists at all into which people can become enculturated. Instead, he points to the fact that people have differential access to knowledge, suggesting that they only need to follow simple rules to go about their business, and are mostly unaware of the existence of any cultural order.

Consequently, he also suggests that we only use "culture" as an adjective. According to Appadurai, such a shift 'moves us into the realm of differences, contrasts, and comparisons' (1996: 12) that, as Borofsky also argues "downplays culture as some innate essence, as some living, material thing" (1994: 245).

Among those who link religion to the evolution of human societies, there is also a widespread tendency to look at religion in much the same way as culture was seen in its classical rendition; that is, as an integrative and harmonious force that united people in communities sharing common beliefs. And since traditional societies are apparently overloaded with myth, magic, and ritual, this is often taken as indication that early humans were also religious. Religion is therefore often depicted as yet another human universal, and it is therefore assumed that its appearance must have coincided with the advent of culture—apparently the two coevolved, perhaps even to the point where they were synonymous. It is only in so-called modern societies that humans have freed themselves from the power of gods and demons.

Rappaport, for instance, contends that religion was a major factor in human evolution. He bases this on the observation that there is no society known to "anthropology or history that is devoid of what reasonable observers would agree is religion." Logically, then, humanity could not have evolved in the absence of religion (Rappaport 1999: 1). Dunbar seconds this opinion, but at the same time he offers us an attempt to explain religion's role in human evolution. He writes: "Religion, in short, probably evolved for bonding social groups and making them work together in the common interest. It was this, above all that relentlessly drove the evolution of the religious mind" (Dunbar 2004: Kindle Loc. 2488).

Wilson also argues that religion appeared in connection with the evolutionary history of humanity (Wilson 2012: 258). For him, religion was a tribal instinct that could only arise in evolution by group selection in a world where tribe competed against tribe. Thus, he finds that religious faith is the logical consequence of the natural dynamism that occurs at this higher level of biological organization (Wilson 2012: 259). Wilson further posits that the core of these religions may be found in their creation myths. He ponders how these fantastic stories came about, and he has a very pragmatic answer to this question: namely, that they are partly based on folk memories of great events that are reworked and ritualized over generations (Wilson 2012: 259). In this account, dreams and myths blend to become religious faith.

Wade (2009), for his part, is a good example of scholars who find that the collective nature of religion gave evolutionary advantages to the groups who practiced it. Wade is very explicit about the integrative nature of religion. He argues that religion embodies the moral rules of a community, and thereby sustains the quality of the social fabric, binding people together for the purpose of collective action (Wade 2009: 2).

According to Wade, religion lent internal cohesion to the groups that practiced it, as well as high morale in times of warfare. As a consequence, more children from these groups would survive, and religion was therefore favored by natural selection (Wade 2009: 12). Because of the survival advantages it offered, practicing religion was written into the neural circuitry of the victorious groups (Wade 2009: 6). Those who could not muster such solidarity died out in harsh struggles between groups.

For Wade "religion is so natural to humanity that it seems to be part of human nature, as if a propensity for a belief in the supernatural were genetically engraved in the human mind" (Wade 2009: 4–5). He even claims that religion served as a primordial form of government for people in the absence of a state. Religion enabled a society to impute its collective wisdom to the gods, and since these early societies had no police and no prisons, religion became almost like an invisible government through its social cohesion and the compliance it commanded (Wade 2009: 15).

But there is at least one obvious problem with these explanations of religion. It becomes an adaptive solution to a later problem. It becomes a kind of evolutionary premonition of a necessary integrative feature that all societies need to function: one that is invisible at the beginning only to become manifest later on. Since early humans could not set up a real government from one day to the next, this construct had to be invented behind the scenes, as it were, beyond the reach of consciousness of those who came to practice it. In this way, human society was born already equipped with an invisible state.

Religion then becomes a complicated solution to a simple problem, for example, keeping groups together. And as authority didn't exist, it therefore had to be invented. But early societies were probably characterized by elder male dominance and a sexist social organization, just as we find today in societies that have been termed "primitive." For instance, the so-called big-men societies in Melanesia are characterized by an egalitarian social organization with no fixed or hereditary status positions, but even these societies are permeated by intergenerational and gendered power relations, and older men dominate both females and the young (e.g., Bergendorff 1996).

All these ways of understanding religion are based on a backward reading of history that imbues the past with today's categories and concerns. In this view, whatever categories we now use, and whatever phenomena we think characterize social life (in the West), are assumed to have been present—or at least latent or dormant—always and everywhere. This stance is known as ethnocentrism, or more specifically in this case, Eurocentrism.

Eurocentrism is a way of describing and explaining other people's customs in terms of contemporary Western categories. And, although Rappaport tries to preempt contenders to his argument that all societies have religion, there are many who have pointed out that this is a Western misconstruction of other societies that are described as if they were mere reflections, or forerunners, of Western societies.

There have been attempts to question the notion of the "primitive world view" as being essentially "religious and symbolic," and that "preliterate people are totally enmeshed in conceptual categories that are essentially religious" (Morris 1987: 1). Alongside this, there has been a widespread tendency to study "tribal religions" in such terms. This has occurred because analysts have disaggregated the religious into different dimensions: "mana, taboo, totemism, magic, shamanism, myth and the sacred" (Morris 1987: 3).

However, as Asad argues, the use of such religious terms assumes that "religion has an autonomous essence" which "invites us to define religion as a transhistorical and transcultural phenomenon" (Asad 2002: 116). His argument is that the concept of the religious—its constituent elements and relationships—is a Western product that changes over time according to historical context. Therefore, religion cannot be made the object of transcultural comparisons (2002: 116). For similar reasons, the use of such universal concepts in ethnographic descriptions has been questioned (e.g., Skorupski 1976; Asad 2002) and, as Strathern (1988) argues, Melanesian ways of life, for instance, have been seriously misread by the imposition of European concepts and prejudice about them.

Thus, the notion of religion has informed our understanding of non-Western cultures, and by extension human evolution, in a rather negative manner.

This understanding is based on a separation of mind and matter which positions "culture" in opposition to something called "nature," a separation that also produces the in-between category of the "supernatural," for example, a world beyond both "culture" and "nature" and which is the realm of the gods. This historic heritage turns all cultures into a variation on a common theme because it relates all practices to a European division of culture, nature, and the supernatural. In the process, non-Western cultures appear inferior and prone to supernatural thought systems.

All these studies assume an explanatory efficacy of Western concepts in our understanding of so-called religious life in non-Western societies which, as Trouillot (1991) argues, has a tendency to reaffirm the exotic character of other cultural systems. But, as Skorupski points out, a ritual statement cannot simply be translated because it would involve the notion of identical meanings expressed in different linguistic forms (1976: 14). Meaning, he argues, is generated in a context and "the overall framework provides the indispensable, culturally shared categories within which a person understands his world" (1976: 227).

Tambiah also cautions us that by "simply naming rituals of non-Western societies as 'magic' and the substances used as 'medicines' and 'drugs,' one cannot thereby attribute to the phenomena so named, by virtue of that naming, characteristic that may be peculiar to one's own contemporary civilisation" (1990: 357).

But this slip-up is still committed by many scholars, because the notions of culture and religion are part of the same conceptual complex that evolved as part of Western history and which therefore belongs to a specific Western ontology. It passes unnoticed by many Western scholars because it is part of their *habitus*. It is their reality. This means that despite all good intentions, the historical heritage of culture and religion has resulted in all people being described according to a common standard—and one which has mostly been unfavorable to non-Western cultures.

The Eurocentric content of such studies went unnoticed as long as the subjects of Western anthropological investigation remained silent. Those days are over. As Ribeiro and Escobar (2006) observe, the spread of anthropology as a scholarly discipline among the world's formerly colonized subject's calls into question the idea of a single or universal anthropology (1). In a world of globalized scholarship, it seems that the translation of a multiplicity of life forms into universalist European categories is problematic (Ribeiro and Escobar 2006: 4).

In this light, the Indian anthropologist Momin (2001) calls for a critical reappraisal of the theoretical and methodological baggage of Western anthropology. According to Momin, such a project requires "investigation into the relevance, and utility of conceptual categories and methodological approaches in the traditions of the non-Western world" (2001: 60). From an

Indian point of view, it seems that the cultures of the former subjects of colonialism and anthropology have "been subjected to a disgraceful generalization" and that anthropologists produced descriptive accounts of quaint customs and beliefs, with the result that traditional thought was "branded as a collection of empty concepts" (Saraswati 2001: 3–4).

We therefore cannot infer anything about early human evolution from present Western categories such as culture or religion. Instead, we have to scrutinize the fundamentals of the human condition—exchange and cultural differences—and see what the ethnographic record about these issues can tell us about the early human condition. In short, we need to trace the origins of culture back to a simple rule of interaction between early humans: one that carried the seed of symbolism.

KINSHIP AND THE CULTURAL FORMULA

Having established the anti-evolutionary nature of traditional conceptualizations of culture and religion, we see that early humans did not just stumble blindly into culture, and neither were they suddenly immersed in religious beliefs. There was no cultural "big bang" in which early *Homo sapiens* suddenly became fully enculturated. Indeed, culture may never have taken the form of an integrated symbolic order. It could not have been invented "wholesale" as traditional understandings would have us believe. And early humans certainly cannot have "invented" a symbolic universe intentionally, or by incrementally enhancing their existing abilities.

Rather, culture must have been a product of new behaviors. Otherwise *Homo sapiens* would not have survived. The first seeds of "culture" must have arisen from new practices. So the development of culture did not require a full, one-off transition into a symbolic universe in all spheres, but merely the appearance of one practice that involved a symbolic dimension. Such a practice had just been invented in Africa, where humans survived by inventing kinship.

Other social scientists have also pointed to the significant role of kinship in human evolution, also pointing to the existence of kinship in every human group. Alan Barnard simply states that kinship is ordered and systematic "in every human society" (Barnard 2011: 85). Barnard finds that anthropology was initiated as a social science when Henry Summer Maine in his *Ancient Law*, first published in 1861, changed the focus in social science from the social contract to kinship, which subsequently became the key focus of anthropological theory (Barnard 2011: 111).

For Barnard, human evolution went through three major stages: a *signifying revolution* in which hominines became capable of using words and therefore also of classifying things, an ability that he finds closely coincides with a very rudimentary kinship system in which spouse sharing was common, in a kind of primitive form of promiscuity (Barnard 2011:135). This stage, Barnard surmises, was followed by a *syntactic revolution* which ushered in categories such as "mother" and possibly "father," "brother," "sister," "son," "daughter" and "mate" (Barnard 2011: 136). Last came what Barnard calls the *symbolic revolution*. In this stage, true kinship coincided with the elementary structure of kinship as defined by Claude Lévi-Strauss involving reciprocity and direct exchange (Barnard 2011: 138). Elementary kinship is based on a positive marriage rule that defines whom you must marry, usually some kind of cross-cousin.

Barnard finds that because the invention of kinship links the nuclear family to in-law groups, who are then also classified according to kin categories, this would provide these people with symbolism because of their continued engagement in exchange and sharing (Barnard 2011: 138–39). For Barnard, as well as for Lévi-Strauss, marriage is communication; it unites kin groups, which tend to create sets of relationships that are both systematic and culturally understood. Therefore, Barnard finds that marriage systems have a "grammar," which calls for categorizations of people in different kin positions which can then be attached to other such categorizations, as in "mother's brother"—a cross-cousin (Barnard 2012: 43). In this way, Barnard finds that marriage and symbolism evolved together.

Bernard Chapais (2008) is also interested in the relationship between kinship and early human evolution, and he also relies heavily on Lévi-Strauss in his attempt to explain human evolution and society. Chapais finds that Lévi-Strauss' "atom of kinship" uncovered a "deep structure" in human society, or what he calls "their lowest common denominator" (Chapais 2008: 301).

This "atom" defines the minimum exchange relationship, which occurs between two groups who exchange spouses. Such a situation produces four kin terms—brother, sister, sister's husband, and sister's child. These are the four positions needed to instigate sister exchange and reciprocal exchange: "brother" as the group that will give their "sisters" to another group, which then becomes "sister's husband group"; and "sister's child" who will repeat this exchange circle. He also finds that human evolution is intimately connected to reciprocal exogamy and that a substantial portion of the exogamy configuration existed well before humans could make use of their symbolic capacity to communicate about their social relationships (Chapais 2008: 302).

Chapais divides this exogamy configuration into three phases. The first phase began just after the *Pan-Homo* split, which produced a society with minimal kinship but with a dormant agnatic kinship structure. Phase two was characterized by hominid groups that were composed of several bi-parental families embedded in a group-wide genealogical web with primitive patrilocality. Supra-group organization was still absent. Then came phase three, in which pair-bonding females acted as peacemakers between their natal kinsmen and their husbands. This peacemaking coincided with affinal brotherhood and set the stage for the brother-sister kinship complex and kinship constraints on marital unions.

So for Chapais, the atom of kinship is not a cultural construct as Lévi-Strauss would otherwise claim. Instead, he finds that the atom has a wholly biological foundation, being composed of three basic bonds with deep evolutionary roots: a kinship bond, a sexual bond, and a parental bond. These bonds together paved the way for the human transition from nature to culture (Chapais 2008: 302–7).

Robin Fox wanted to convince us that humans and the human mind are a product of evolutionary forces, and that humans therefore never did escape nature, as others would have us believe. He argues that humans act in terms of categories and concepts that they impose on nature (1979: 138). To do so, Fox searches for the basic feature of human life, and like Barnard and Chapais he finds that Lévi-Strauss is useful because Lévi-Strauss writes about the "fundamental structures of the human mind." These fundamental structures are: the exigency of the rule as a rule; the notion of reciprocity as the most immediate form of integrating the opposition of the self and others; the synthetic nature of the gift, and the transfer of valuables that turns these individuals into partners and adds new qualities to the transferred object. Humans are, then, characterized by rules, reciprocity, and gifts, and humans universally make rules, classify, and exchange the things they makes rules about, thereby setting up relationships (Fox 1979: 140).

To account for the evolution of these abilities, Fox looks at changes in the relationship between humans and their environment, which he thinks changed rather slowly over a long stretch of time. He sees this as involving a twin set of changes consisting of adaptive problems *vis-a-vis* the environment and the organizational changes these caused. But the main event, he notes, was the advent of tools, which, for Fox, suggests handedness, the lateralization of functions, cooperation, and speech in some rudimentary form. But the essential thing that these tools made possible was the systematic introduction of meat into the diet, which according to Fox called for an organizational form that could acquire and distribute it (1979: 144).

He also finds that this new situation changed the relationship between the sexes due to the new division of tasks, thereby also changing the power relations in the groups and the strategies used to become powerful. Instead of dominating young males to get access to females, the new strategy was to dominate in order to use the females to make alliances (1979: 147). The resulting systematic exogamy cannot exist without identifying "ours" and "theirs," which gave rise to new rules of taxonomy and symbolism. Here again, kinship is invoked as the core explanatory factor in the human symbolic revolution, and is therefore regarded as a pan-human phenomenon: "and whatever else humans classify or not, there is no known human group that *does not classify its universe of kin*" (italic in original 1979: 149).

All three authors point to the importance of kinship in the evolution of the human lineage, and they also find that it was kinship that enabled humans to transcend the animal kingdom and become intelligent and live a symbolically constituted life. But both Barnard and Chapais find that this transition was part of a general evolutionary process that pushed the human lineage forward toward ever greater complexity. They were practically destined to become *Homo sapiens* . From a purely evolutionary point of view, however, humans lack an adaptive behavioral transformation that would produce kinship. Complex living in itself does not necessitate marriage, and symbolic life could in principle also have arisen without kinship. So there have been an environmental situation to which kinship was an adaptive response.

Fox comes close to depicting such a situation. He finds that tool-making favored cooperative living, instigating a rudimentary division of labor that made alliances preferable, since women had to mediate between otherwise aggressive males by marrying them. Tool-making is, in his account, the medium through which the brain was challenged, leading to all the other transformations including the advent of kinship. This account was plausible at the time when Fox wrote his thesis about human evolution, but unfortunately has now been invalidated. We now know that new technologies were invented after the speciation process, that hominine used the same technologies for long stretches of time, and that new species used old tools for a while before changing them.

These studies all claim that kinship was important in the final evolution of *Homo sapiens* . But none of them have pointed to the adaptive role played by kinship, which was premised on exchange between groups. This new invention was crucial for the survival of the species. Thus, when groups were slowly drawn apart they would have had to continue to recognize their kinship ties in order to continue exchanging resources. They would therefore have had an incentive to maintain their relationships, and to do so they would have to keep track of their kin. By formalizing the exchange of spouses, they would have been able to form alliances and keep up those relationships.

Alliance through marriage would therefore have been inseparable from exchange, and exchange relations were based on kinship. That was the only option for early humans. Without kinship they would just have been groups living in similar and parallel environments, just like *ergaster*. Kinship enabled them to formalize their relationships to other groups through the only means possible: human relations. Having forged these alliances, instead of just being new additions to a group without any lasting relationship to their groups of origin, the incoming women would have known which group they came from and thereby also their decent. By knowing their descent they became wives who *represented* their natal group. Women thereby became the primordial sign. Being forced to share resources between groups was what led the early humans to invent an elementary form of culture.

And that is not all. Early exchange contained another symbolic element. The need to survive was what triggered the first forms of exchange: different products were exchanged for one another. In the Cape Floral Region we saw that shellfish were probably exchanged for tubers. This also enabled comparisons between "different" things. By comparing different things through exchange, *Homo sapiens* engaged in a symbolic act where the primordial sign was extended to cover other phenomena.

As Lévi-Strauss puts it: "The function of signs is, precisely, to express the one by means of the other" (1979: 14). By now, *Homo sapiens* had unintentionally constituted a symbolic universe by making alliances to enable survival in harsh conditions. They had invented the sign in order to exchange, thus simultaneously inventing symbolism. By acting persistently according to these new modes of behavior, kinship, and alliances, they socialized themselves increasingly into an emergent, symbolically constituted universe.

To understand the nature of this transition we have to rely on anthropological theory. This is where the study of kinship by Lévi-Strauss is particularly relevant. He sees kinship as based on relationships between groups, not as something that singular groups practice. And he is concerned with how this relationship is established. For Lévi-Strauss, exchange and exogamy were linked phenomena, as he wrote "exchange—and consequently the rule of exogamy which expresses it—has in itself a social value. It provides the means of binding men together" (Lévi-Strauss 1969: 480). Women-mediated partnership is, thus, the very foundation of kinship.

Lévi-Strauss himself argued against any evolutionary interpretations of his analysis, writing that "we should resist to the very end any historic-geographical interpretation which would make restricted or generalized exchange the discovery of some particular culture, or some stage in human development" (Lévi-Strauss 1969: 463).

Still, his theory identified exogamy both as the core organizational principle in contemporary human society as well as in its archaic manifestations. And in relation to the time and place, we now know that the original human society—and kinship—emerged in Africa, some 200,000 years ago. Unwittingly, and probably also without wishing to, Lévi Strauss's *Elementary structures of kinship* gave us the key to understanding the transition into *Homo sapiens* as well as to the elementary structures of culture—or rather its algorithm, which charted relationships between two or more groups.

According to Lévi-Strauss, marriage is the primordial act of exchange, linking two groups together. By initiating marriage, *sapiens* also instigated kinship: as Lévi-Strauss argues, kinship can only be allowed to be established and perpetuate itself through specific forms of marriage (1968:

50–51), for example, a specific relationship between two groups. Lévi-Strauss cites Radcliffe-Brown, who claims that the essence of human kinship is to establish relations between elementary families—but in fact it is not the families that are elementary but the *relations* among them (Lévi-Strauss 1968: 51).

These relationships among groups led Lévi-Strauss to claim that the presence of the maternal uncle is a necessary precondition for a kinship structure to exist. The uncle, so to speak, represents the link between groups. Thus, for kinship to exist, three types of family relationship must always be present: a relation of consanguinity, a relation of affinity, and a relation of descent (Lévi-Strauss 1968: 46). This is why the relationship between nephew and maternal uncle has been the subject of considerable anthropological debate with regard to so-called primitive societies.

According to Lévi-Strauss, the avuncular relationship is always part of the elementary unit of kinship, which consists of a man, his wife, and the uncle who gave the female away from another group. In a sense, this triad forms the true atom of kinship (1968: 48) because it defines the group that one's own group is allied to. In its most abstract form, this structure introduces a *difference* that can only be maintained through a relationship—the wife is different because she maintains her relationship with her group of origin. Otherwise an exchanged female would just be seen as yet another woman who is new to the group. But by introducing a relationship, that female becomes a wife—she is, so to speak, given *meaning*. In other words, she becomes a sign that *represents* her group of origin. Thus, the wife may be said to be the primordial sign. This uncle-man-wife "kinship atom" is a universal structure that always and everywhere defines the relationship between human groups, giving rise to a system of social reproduction that produces differences while maintaining relationships.

But Lévi-Strauss also calls this an elementary structure, because it is not directly observable in the terms people have used to designate their kinship relations. To him, kinship systems are built by the mind on the level of unconscious thought (1968: 34). Thus, he finds that kinship shares the same characteristics as phonemes in languages, which are elements of meaning, and he argues that kinship systems do not, in fact, consist of objective ties of descent or consanguinity between individuals. Rather, for Lévi-Strauss, kinship is an arbitrary system of representations (1968: 50). Thus, although kinship may represent lasting relationships between groups, to Lévi-Strauss it is strictly about symbolism (1968: 51). This is so because kinship constitutes a lasting relationship between groups that is independent of the personnel occupying the various kinship positions—they merely represent a given symbolic relationship.

The invention of kinship is, therefore, a key moment of transition in human evolution. By inventing kinship, early *Homo sapiens* made the initial transition into a symbolically constituted universe, and kinship terms may, therefore, have been the first phenomena that needed some kind of explicit articulation. Kinship positions would have had to be articulated using different sounds—or words.

Lévi-Strauss thinks that "primitive" societies are based on what he calls mythical thought, but that this is just as rational as "modern" thought, since both forms share the same basic premises. It is only the subject matter that they address that is different. To Lévi-Strauss, there are some basic logical processes at the root of mythical thought. These always progress from the awareness of oppositions toward their resolution (1968: 224). This is exactly the situation that is involved in exchange between groups. Furthermore, if these opposite terms are assigned to the things that are exchanged, they will always require a third term to mediate between them: "then one of the initial polar terms and the mediator can be replaced by a new triad, and so on" (Lévi-Strauss 1968: 224).

This can be seen as the basic structure of symbolic thought, or what we might term the elementary unit of meaning that corresponds to the basic unit of kinship, for example, an opposition plus a relationship that together produce a third term that combines the first two. This is actually just a description of the relationship between the first modern human groups based on the way they adapted to their new environment. And so, it seems that what we now call culture is *a byproduct of niches that are connected through kinship.* This was a symbolic process, but it was produced by a new survival practice—the kinship structure—to which it will always remain connected.

These oppositions and their mediations that can continually be replaced by new triads are basically metaphoric in character. This is, then, the basic structure of culture: a kind of algorithm that can be endlessly extended metaphorically to cover an infinite range of new situations and products. Lakoff and Johnson argue that this system provides us with *Metaphors We Live By* (1980). Along similar lines, an important discovery of cognitive science is that while the conceptual system used in the world's languages makes use of a relatively small number of basic image schemas, the range of complex relations that can be built out of these schemas is enormous (Lakoff and Johnson 1999: 35–36).

Thus, things can be expressed in terms of ever-expanding metaphors that include more and more things. This would mean, as Lakoff and Johnson also argue, that categorization is not a purely intellectual matter; rather, the formation and use of categories results from experience (1999: 19). Accordingly, we categorize things the way we do because we interact with the world in the way that we do.

So, it becomes clear that human categorization is organized along the same lines proposed by prototype semantics. Within this scientific tradition, Lakoff and Johnson suggest that people understand one domain of experience in terms of another, and that understanding takes place in terms of entire domains of experiences and not in terms of isolated concepts (1980: 117). They also argue that some concepts emerge directly from experience and that these are related to emergent metaphors that carry these direct experiences over to less clearly delineated experiences (1980: 58). Symbolism is, therefore, linked to experience and to people's ways of being-in-the-world.

Concepts and categories are thus formed as directly emergent properties of interaction in and with an environment, in which some of the properties are produced by the interaction. This is the cultural formula: an elementary unit of meaning extended indefinitely by metaphors to other domains of life. Things become linked by means of metaphors to form very loosely defined categories, in which some things are more central to that category than others, and these things then form prototypes for that category. Taken as a totality, these categories offer a very flexible cultural order that is always "in the making," producing a highly malleable symbolic system.

The emergence of symbolism can therefore be seen as a result of following a set of simple rules that connect groups to each other. Once this symbolic universe was established, there was no way back for early humans. The transition into a symbolic universe immediately gave early *Homo sapiens* an adaptive advantage by expanding their biological niche.

That humans started to use abstract categories is clearly reflected in the archeological records. At the same time as the appearance of *Homo sapiens* we see remarkable changes in these records. As the Committee on the Earth System notes, many behaviors that are usually attributed to *Homo sapiens* have left traces in the archaeological records from about 200 to 100 ka, notably the use of ochre and the long distance exchange of material and bead ornaments (2010: 21–22). Seen from the perspective described above, this is exactly what we would expect.

It is commonly believed that ochre and beads were used as bodily decorations, and therefore that these finds indicate symbolic capacities (Cela-Conde and Ayala 2007: 338; Wade 2006: 32). At the same time, it is evident that these items had been transported over long distances, probably in some kind of relay trade, from group to group (Oppenheimer 2003: 127). But to get a better understanding of the purpose of these shells and ochre, we need to return to the ethnographic record to see what these things were actually used for in so-called traditional societies, notably in those places where the most direct descendants from early *Homo sapiens* lived, for example, Australia and Papua New Guinea, the eldest and next-eldest lineages (Stone, et al. 2007: 175).

Items akin to beads and ochre have probably been use by human groups ever since early human times. The traditions surrounding use of these items in Australia and Papua New Guinea are very complex, but we may assume that the items have the same basic symbolic functions today as they always have. In Papua New Guinea, shells, and beads, along with other items such as sea shells, tortoise shells, and birds' feathers, are generally made into decorative items, for example, used as arm bands or strung together as necklaces (Bergendorff 1993; 1996b).

These valuables were also used for personal decoration at ceremonies and as payment for different services performed by important persons such as chiefs and sorcerers. But they were first and foremost used as bride prices paid between clans for the procurement of women. These symbolic items were conceptualized as valuables and without access to such valuables people often found it difficult to get married. As some of the first explorers and priests in the Mekeo territory of Papua New Guinea noticed just after the turn of the twentieth century, bride price was raised by the whole family, both male and female lines, and the purchase depended on the financial situation of the parties involved (Egidi 1912: 9). Some were too poor to get married, or had to settle for a widower or an excommunicate (Kowald 1893, Egidi 1912).

The Mekeo, like many other ethnic groups in Papua New Guinea, also marked their marriageable women in a very permanent way by tattooing their entire upper bodies in clan-specific designs. In this way, everybody could see where they came from when a woman left for her husband's clan (Bergendorff 1993; 1996).

Valuables and body-marks can, thus, be seen as symbolic expressions of the two predominant kinship relationships: alliances and descent. Marriage, or alliance, is symbolized using "valuables" and descent is marked with tattoos that could easily have replaced an earlier form of marking using a substance such as ochre (Nettle and Dunbar 1997).

In the light of the data from Papua New Guinea, it is reasonable to think that the use of beads and ochre in Africa might be the first symbolic expression of the new kinship networks based on alliances between groups, since they connected descent and marriage: beads symbolized the matrilineal alliance relationship between groups, and ochre served as a marker of patrilineal descent.

In this light, alliance and exchange should be seen as the biggest adaptive advantages ever invented by the human population. Exchange makes it possible for humans to live in a wide variety of "ecological niches," including those that do not otherwise provide a living for them. Exchange is, then, the most fundamental property of human life (Ofek 2001): the key feature that distinguishes human beings from all other living beings.

All other animals live in ecological niches in which they adapt, evolve, and become extinct. Humans transcend this predicament through exchange. Fredrik Barth once carried out a study of ethnicity along these lines. He found that there was a positive bond that connected several ethnic groups in an encompassing social system that depended on the complementarities of the groups, and that such groups provided important goods and services for each other. In Swat, Pakistan, three different ethnic groups existed together exploiting different environmental zones and they exchanged specialized products with each other, integrating irrigation farmers, scattered pastoralists and mixed herder-framers into a common system (Barth 1969: 17–19).

So exchange probably enabled early humans to specialize in one form of product, and to trade this for another product in order to survive. Thus, human populations form networks of interacting groups that are able to transcend the limitations of ecological niches by exploiting symbolically constituted differences between groups, who then trade with each other. *Homo sapiens* had, by now, become the one and only trading species. This breakthrough was enabled by kinship, which carried the seed of symbolism, expressed in the exchange of females, who went from being random contributions to their new groups, to being wives *representing* their groups of origin.

CULTURE IN THE MAKING

Kinship produced symbolism, and kinship is about alliances between groups, and alliances between groups are about managing resources so that single groups are no longer dependent on their own local niche to provide all their necessary resources. We might therefore also surmise that "culture" at its most basic level is about relationships, and not about the inherent qualities of single groups, as cultural relativism would have us believe.

This angle on culture might also prove a better way of explaining cultural differences, since living in networks produces different life conditions in its different locations, which must be negotiated symbolically. And since such networks are never in equilibrium and therefore continually changing the local living conditions, this might also be a better way of explaining cultural change, which is almost impossible seen from the perspective both of cultural holism and relativism.

And if it is true that this form of culture gave early humans an adaptive advantage, we may logically assume that it is still with us. So, although cultural orders give the superficial impression of being integrated, we should shift our focus from the "order" to the qualities of the networks that all social groups and societies belong to.

The advent of culture should therefore be seen as part of the adaptive process that caused our speciation into *Homo sapiens*. Seen from this perspective, culture is not a detached sphere with a life of its own, which randomly shaped the modern human condition according to its own rules. It was, rather, an adaptive process which enabled humans to shift from foraging in isolated groups to becoming integrated in reproductive networks.

The survival of species that live in isolated foraging groups is entirely dependent on the immediate environment. Each group is intimately connected to its environment, and the groups wax and wane with the conditions within their range. Groups in such a situation engage in a reciprocal relationship and an immediate exchange with the environment. There is a direct *link* between the group and its environment.

But humans were no longer organized in isolated foraging groups. By inventing kinship they had become integrated in networks of reproduction. This change completely altered their survival situation compared to all other animals and to their precursors. Living in networked groups means that even if some of the preconditions for survival are not present in the immediate environment, they are available through the network. Survival is, then, to a great extent dependent on the *relationships* in the network.

Living in networks produces new properties in the reproduction of the single groups that constitute it, properties that lie beyond its members' direct experiences. In networks it is not only the foraging strategies of single groups that matter for their survival, but the combined character of the network and the available resources within its entire range. These two factors together determine the quality of life of the involved groups.

Being a node in an exchange network is qualitatively different from being an isolated group. Being part of a network means that groups and individuals depend on other groups and individuals that are not physically present. In a network, the conditions of social reproduction for each single group are, therefore, beyond their immediate control. When people start to live in networks they become part of a social reality that is beyond their direct experience but which nevertheless influences the conditions of their life locally. Something external and imperceptible is introduced that becomes a constant factor in local life. In other words, networks produce *emergent properties* in the social fabric, properties that are felt to be part of "reality" and the constitution of the world but which are beyond immediate sensation.

Such emergent properties are felt to be real forces, and they inform local ontology. But they cannot be measured or experienced by outsiders conditioned to live with other emergent properties. Groups living at different nodes in the network automatically become different. Their lives are framed by different emergent properties, and thus they also become culturally different. This state of affairs is a great symbolic enhancer. It is the mechanism that creates cultural diversity.

Exchange in itself is a symbolic process. Exchange involves comparisons of differences, and assessments of the value of one item in relation to another. Therefore, exchange is basically cultural. There can be no exchange without culture, and there can be no culture without exchange. Exchange and culture are two complementary sides of alliances between different groups who live in networks. From an evolutionary point of view, exchange and culture are negotiations of the relationship between groups. With kinship, exchange and culture became the two elementary conditions of human life. Kinship is at once material and symbolic, involving differences and alliances.

Cultural differences result from groups being located at different nodes of a network. So what of religion? That which we call "religious" arises naturally from the experience of living in a situation where not all properties are generated by the immediate environment. Rather, some properties stem from the character of the networks that link its groups together. Religion can, therefore, be seen as an expression of living in a networked social universe where something beyond one's immediate experience exerts a great influence over one's life. But it is doubtful whether in its early days religion had the same connotations which it does today.

This must have been a brave new world for our early humans. Their situation posed immense practical and intellectual challenges that required cooperation, coordination, foresight, insight, and memory. Because of these challenges, exchange conditions, social properties, and cultural elaboration become part of the same social package. These are, therefore, not separate social phenomena that can be analyzed independently.

Chapter Five

Local Strategies and Culture: The Mekeo of Papua New Guinea

Culture may therefore be said to be an expression of local survival strategies. An example is needed at this point to clarify how "culture" is linked to positions in a network, and that exchange, social condition, and cultural elaboration are intertwined. Once again, Papua New Guinea comes to mind because it features a huge variety of different cultural groups all contained within a rather small area. Some of these groups are relatively small and are therefore also well-suited to this kind of investigation.

This time, we will take a look at the Mekeo people who live in Central Province some 150 km away from the capital, Port Moresby. At the time of investigation, the Mekeo numbered around 12,000.[1] My example is conducted from the perspective of global anthropology which seeks to go beyond Western ontology and rejects any description of local practices using preconceived ideas or categories. From this perspective, one cannot talk of "rituals," "myths" or "magic." Instead, we have to search for simple survival strategies, chart their linkages to the social environment, and explore how all this is symbolically negotiated (Bergendorff 2005; 2007; 2008; 2009a; 2009b).

Survival strategies matter because that is how everything started, and where the primordial act of symbolism originated. Our task is not to elucidate cultural wholes and how these shape practices, but rather to understand the strategies that people pursue locally, the kinds of exchange relationships they engage in with their environment, and how their interactions generate properties in the social order. Finally, we must investigate how all these elements are symbolically articulated.

It is local survival strategies and their conditions that are crucial to understanding how the social life of the Mekeo is organized (Bergendorff 2003;

2007; 2008; 2009b). From the historical documents of the mission and government it is clear that well into the colonial era, a clan needed both land and valuables to be able to manifest itself as a localized clan. To survive socially, a Mekeo clan could follow two simple rules: secure land, and find marriage partners for their children; land to produce a food crop for the clan to survive; marriage to secure the continuation of the clan. Although complementary to the continued existence of the clan, these two rules had different underlying principles producing properties in the social order that were different from the rules themselves.

The first of these rules, securing land, could be pursued through different strategies, for example, inheritance and territorial expansion. Seligman describes how this translates into a difference between older chiefs, *lopia fa'a*, and younger chiefs, *lopia ekei*. Fission occurred when a strong group was divided into two groups called, respectively, *faangiau* or "first born" and *ekei* or "subsequently born," and the latter established itself on new land (Seligman 1910: 337). As a result of these strategies, older lineages consisted of landowners; while younger lineages went in search of new land.

In a limitless space, this strategy with its resulting fission process would pose no problem. However, the limited size of the fertile Angabanga Plain, combined with the swidden agriculture practiced by the Mekeo, complicated matters. As Governor MacGregor observed, "the whole country has at some time or other been cultivated. It contains no old forest whatever, until the hills are reached" (1890: 80). So, in reality, land acquisition was no easy matter, and land possession became an issue of contention between elder and younger lineages. The latter were progressively deprived of land, spurring them to abandon their parent clans and go in search of new land (Seligman 1910: 336–41, Bergendorff 1996: 27). In this way, the unintended consequence of inheritance—its emergent property—was continual land disputes and attempts at conquest.

The second simple rule of social survival was to find a marriage partner for one's children. One widespread strategy was to pay for a wife. But all parties to the marriage wanted their share of the valuables given. We know that well into the twentieth century the Mekeo exchanged bride price in pairs: the groom's clan paid bride price to the bride's father's clan; while the groom's mother's clan paid bride price to the bride's mother's clan, a rather expensive undertaking.

Paying bride price would pose no problem if valuables had flowed freely into the Mekeo area. But they did not, which is clear from the description made both by government officials and missionaries. As Father Egidi observed, just after the turn of the twentieth century, bride price was raised by the whole family, both male and female lines, and the purchase depended on the financial situation of the parties involved (1912: 9). Some were even too poor to get married, or had to settle for a widower or an expelled person

(Kowald 1893; Egidi 1912). And the price paid also depended on the distance to the coast and "as a rule more has to be given on the coast than inland" (Kowald 1893:62).

Nonetheless, paying bride price in a situation of relative scarcity enabled larger groups to fortify their positions by absorbing members from other clans or even whole clans who could not otherwise meet their obligations to pay bride price. The bride price strategy thus created a tendency toward the fusion of clans in which poorer groups lived in fear of being taken over by their in-laws. So the practice of paying bride price had the unintended consequence, or emergent property, of creating status differences between clans, favoring the bigger clans while at the same time countering the fission process resulting from land acquisition by establishing alliances, or by fusing sections of weaker clans with stronger ones.

Seligman describes the net result of these two simple strategies: land acquisition and marriage: fission and fusion. Writing about the history of a small community called Afai, he notices:

> The number of times that portions of clans have come together and after having seemingly coalesced so as to form a village, have separated to fuse again partially before once more wandering apart. (Seligman 1910: 315)

Accordingly, these two "simple" rules of social survival, land acquisition and marriage, informed the social condition in the villages by produced properties in the social fabric that were a combination of the social characteristics both. Land acquisition produced a *difference* between older and younger lineages, while marriage produced a situation of four exchange *partners*, together producing a social reality based on a dualism in which each part was subdivided into two parts (Bergendorff 1996: 52). These properties were articulated in the prototype of clan constitution as four positions of an optimal—but not ever-present—institutional order split into two sections: *lopia* (chief)—*ungaunga* (sorcerer); and *iso* (war-chief)—*faia* (war-magician), living on either side of the village plaza.

Thus, the two strategies of acquiring land and valuables for marriage in order to secure the existence and continuity of a clan shaped the social condition by producing a dualism in the Mekeo social world, in which each side of the clan had two properties. However, these two strategies also shared a common denominator. In the first strategy, inheritance determined the amount of land accruing to a given group, favoring the older lines of descent and thus equating ancestry with social position. In the second strategy, to maximize the clan's partners—its access both to valuables and to a workforce—previous alliances determined the potentially marriageable clans, turning alliances into a question of descent and ancestry. Accordingly, both strategies meant that land and alliances were linked up with inheritance, and consequently position and status were to a large extent determined by ancestry.

In this way, one's ancestry was manifest in a multitude of different actions: in marriage, land issues, warfare, status, and position, irrespective of how these practices were actually carried out. Accordingly, ancestry, or the concept of "ancestor," emerged as a set of properties that occurred naturally in the daily lives of the Mekeo. It was a "directly emergent concept" basic to the Mekeo conceptual system and produced through interactions with their social and natural environment (see Lakoff and Johnson 1980: 69–76).

Seen this way, the category of "ancestor" was not defined by its distinctiveness and inherent qualities with necessary and self-sufficient features (Taylor 1989: 23). Rather, it is best understood "as an experiential gestalt": a prototype composed of a cluster of elements produced by the two interlinked strategies aimed at the clans' survival (see Lakoff and Johnson 1980: 70). But as a gestalt, this cluster formed a whole that was more basic than the parts: it appeared as though social "reality" was influenced by ancestors.

Ancestors were an *emergent property* of the exchange network. Thus, ancestors were conceptualized as part of existence, a real force living in an unseen dimension of life from where they wielded great influence over the living. So, ancestors were part of Mekeo ontology, their way of perceiving the constitution of the world. For them it had nothing to do with religion. Furthermore, there is nothing integrative about living with your ancestors in this way; they just represent previous struggles for positions in the network.

More generally, we can say that in a networked society such as it was constituted in Mekeo society, the ancestors, so to speak, shaped the social conditions of the living. They were experienced as an active force in the alliance networks. They were also the ones who instigated the alliances that the present generation had to negotiate. Being a position in a network means that people are connected to people beyond the horizon—it matters less if they are dead or alive; they are still the ones who have influence over your present condition.

Having invented kinship means that the dead still represents the relations between groups, they are still a kind of active agent in present alliances, without them there would be no networks. As Barnard argues "afterlife can be seen as a byproduct of alliances . . . the past in the present" (2011 Kindle Loc. 2340). Any person is linked to their network through the dead. By burying them, they retain their role.

However, humans are not just instinctually programed to interact in a certain way with their environment in a way that influences the condition of social life by producing social properties; these emergent properties are perceived by people and subsequently negotiated, as in the case of "ancestors."

In the early days of colonialism, Mekeo emergent properties formed a dualism in which each part was subdivided into two parts. These distinctions were manifest, among other things, in stories that are told in the villages about the Mekeo cultural heroes, A'aisa and Isapini; about the genesis of the institutional order and other significant events in their history. And yet there are many stories that elaborate on similar themes, and everybody seems to have their own version. Such stories tend by anthropologists to be labelled "myths" because "anything is likely to happen" (Lévi-Strauss 1963: 208).

But in these narratives we can see how the Mekeo interpret their emergent properties and thus how they understand the relationships that organize their social universe and the principles they think govern their substances and appearances. So, here we will treat them as the Mekeo do: as stories about real issues.

One such story is about how A'aisa gave death to the Mekeo people.

> At one time, A'aisa gathered all the people, telling them he wanted to introduce them to death. A'aisa needed a volunteer but all refused because they did not know about death. He then told the people that death was like sleeping. Finally, a man agreed, and A'aisa told him to lie down and sleep. After a while, A'aisa woke him up and they sat together. Then, A'aisa asked the man to sleep again. This time he was dead. A'aisa wrapped the body in a piece of tapa cloth and carried it to the top of the mountain O'opo. From here, he told all the people to gather at the bottom and be ready to catch Imala, the dead man, when he threw him down. But the people became afraid when they saw the falling body and stepped back. A'aisa then informed them that if they had caught the falling body they would have had eternal lives, but since they had mistrusted him, they would be gone for good when they died. After having introduced death, A'aisa cast down the institutional powers, making one man responsible for each. These men then became *iso* (the war-chief), *faia* (the war-magician), *lopia* (the peace-chief) and *ungaunga* (the sorcerer).

Here we have a story about the separation of existence into life and death and the origin of the four leadership positions or qualities, providing a narrative explanation of a duality in which ancestors have lives that complement those of the living. But these two forms of life were not two completely separate domains. They have a common sphere: "sleep." Death is like sleeping. It is another state of being. Thus, the living can "see" their ancestors, and intermingle with them when they sleep. Thus, the Mekeo lived both among the living and the dead, and sleep was the meeting ground between these two spheres of being. In this way, ancestors were explained as an active force in everyday life, and their existence was proven by relating them to "observable facts" such as life, death, and sleep.

Thus, the emergent properties were retold as though existence were divided between two spheres of life, and as though ancestors had an active role in the lives of the living. Subsequently the nature of these domains became a focus for speculation, as in one story about Polepo who, with A'aisa's help, went to fetch his wife in Aunipangua, literally "the village of the dead" and brought her back to the realm of the living.

> One day, A'aisa passed by Polepo's place on his way to "a far-away land." Polepo persuaded A'aisa to take him on his journey. Every night, A'aisa went off into the bush, returning in the shape of different animals, and tormented Polepo, who persevered in his ordeals until they arrived at the village of the dead, where they finally located Mapo.
>
> Polepo convinced his wife to return with him to "the land of the living."
>
> Hereafter, the couple thanked A'aisa for establishing the contact between the living and the dead and started on their journey back to their village. Once settled, they planted a new garden. When the banana plants began to sprout, Mapo went out to cover the young bananas with leaves, but she dropped some of them. Polepo picked them up and handed them to her as she stood on a ladder, but to his horror, he discovered that he could see right through her, through her vagina and up to the sky. After this unpleasant incident Mapo felt ashamed and compelled to return to Aunipangua.

In this story, the relationship between the two realms of existence, life and death, is recounted as a story of appearance and substance and the relationship between the two. When Mapo returns to the land of the living she is only appearance. She is hollow, her substance remains in the land of the dead. So the two realms—life and death—relate to each other as appearance and substance, respectively.

We also learn that living in the two realms is similar, but when Mapo returns to the land of the living she is only appearance. She is hollow; without substance, indicating that the two realms are constituted differently as one of appearances and one of substance while both contains the "same" forms of living consisting of a cover and a substance. However since one realm is based on appearance and the other is based on substance, but still containing the same life forms, they are inverse realms.

A third story, also about A'aisa, addresses this topic of matter changing form. After a series of misunderstandings and fruitless exchanges between A'aisa and his brother Isapini, they finally killed each other's sons.

> When the son of A'aisa died, A'aisa buried him at Avungovungo. Isapini's son is also buried there. A'aisa was distraught with sorrow, so he dug up the corpse of his son and started travelling with it. He went to several different places, weeping all the way. He came to a stream where he washed his son. From there, he went to Ikonga, where the navel of the corpse burst open. He went on until he reached Ko'oko'o, where he found a place to bury his son's navel. Then, he travelled on to Koana, then to Kekenina, and on to Kaliko. When he arrived there his son's corpse was almost completely rotten, and he buried him there. A'aisa then went to Bereina and threw a big feast, after which he went back to O'opo. The bones stayed where A'aisa buried his son, while he took his son's *laulau* and carried it to the mountain.

This story tells how A'aisa carries the rotting body of his son—named Isapini after A'aisa's brother, who killed him—around in the Mekeo area, while pus is dripping onto the ground. An important point that is explicit in the story, but was explained to this author by an *ungaunga*, is that in each place that this pus drips, it grows up as a plant that is thought to mirror the body part from which the liquid came. To the Mekeo, these plants *are* actually A'aisa's body parts—they share identity, or substances, with him in spite of their different appearances.

From these stories we can discern how the Mekeo formerly interpreted the emergent properties stemming from living in a networked society, and by extension the forces which they believed operated in the world to produce this situation. For instance, to explain "ancestor," the Mekeo divided existence into two permeable domains, one of appearances/forms and one of hidden substances/essence. To account for the relationships between these two domains, they found that the world of hidden substances protruded into the visible world in different shapes that were given their form and identity by their relationships in the invisible world of the ancestors. These basic ideas formed the common framework of Mekeo discussions, negotiations and practices.

All these stories can, thus, be seen as attempts to make sense of emergent properties by "translating" them into narratives that explain the hidden principles and properties that seem to coexist in the world along with "obvious" properties such as night and day, up and down, and so on. This has consequences for the way elements are supposed to interact. These elements, or the two realms, form connected wholes in which everything has a two-sided existence.

Since emergent properties are inscribed into the environment and even into the constitution of the human being and the understanding of personhood, they also inform the logical connotations of practices, although not their actual execution. In this way, emergent properties inform the logic of the appropriate and the possible as a kind of praxis-space, that tells us something about local ontology with its "laws of motion" in the social universe and the relationships between elements, that together produce the conditions for the elementary forms of interaction. In this way, what is considered "appropriate" is based on an often tacit understanding of "what is true, reasonable, natural, right, and good . . . but it does not guarantee technical efficiency or moral acceptability" (March and Olsen 2004:4).

From the stories above we see how the emergent gestalt of ancestor was metaphorically extended to all things and beings that were subsequently seen as existing in two realms simultaneously. In this way, they derived their identity/appearance from their relationships in the invisible world. More generally, we have a prototype of existence so that every kind of being consisted of an epidermis, a substance, and the hidden relationships between them.

Plants, animals and humans could thus be described in the same terms as existences, living at the intersection of the two realms. This idea was metaphorically extended to become cosmic principles, which were exemplified by A'aisa and his brother Isapini.

In this way, emergent properties are inscribed in the environment; they are given life through their metaphorical connections to more tangible phenomena, and through their relationships with aspects of the human constitution. As such, emergent properties become real features of social life, not only invisible features, since they are inscribed as the conditions for life itself. What recent ethnographers have seen as religion is actually part of a social reality based on the way the Mekeo life-world was constituted and reproduced.

There is nothing inherent in this Mekeo ontology that we can call an enduring Mekeo culture that is particular to these people and which will survive eternally. And in fact it did change, rather abruptly, in around 1990. At this time, the options for exchange had changed rather radically. Now the villagers earn their living mostly from money income earned by selling betel nuts to buyers in the capital, Port Moresby. This is a completely different survival strategy than the previous one, and it carries some new emergent properties that the villagers presently are struggling to make sense of.

It follows that the ontology in one part of an exchange network is not transportable to another position in a trade network, and certainly not between different networks—even though most networks are connected these days. Thus, categories cannot have universal contents. Every social system creates its own reality, linked to a position in a network and the survival strategies possible in the place where it is located.

In every position, people pursue their livelihoods within a given framework, and at any given time in a given position there are probably few options, and these will tend to maintain the system—for example, what we have called the cultural order. But if the cultural order means nothing by itself but constitutes an integrated part of exchange, any small deviation in survival strategies will have to be symbolically negotiated. So no group is permanent. To study "the social" we therefore need to establish the character of the network and the nature of the emergent properties in its different positions.

By inventing kinship, humans came to live in networks of reproduction. This completely changed the human condition. This insight also enables us to explain the two predominant phenomena of social sciences; economy and culture or cultural differences. These phenomena are linked in networks that produce different emergent properties in their different positions. Had no networks existed, humans would still have formed different but similarly constituted groups in different localities. But networks provide a reproductive suprasphere that adds an extra layer of properties to mere survival. Networks create differences as well as linkages.

Unfortunately, Western social science has separated the relationships (or the network) from the groups (or nodes), which are then analyzed separately. But they are not separate social spheres, each with their own dynamic that can be studied in its own right. This is the illusion of a particularly Western ontology that has separated the social into spheres such as state, civil society, and economy, or culture versus economy, and then struggles to unite them again. Each of these artificial spheres is then studied by its own specialists: economy by economists, culture by anthropologists. And each has developed its own very different theoretical approaches and models to deal with its supposed corner of reality. But in actual fact, these spheres are but complementary aspects of life in a networked world.

NOTE

1. This chapter is based on my own fieldwork among the Mekeo people, which started in 1989. Since then, I have conducted six periods of fieldwork, the longest being one year for my PhD dissertation. Altogether, I have conducted more than two years of fieldwork here, studying very diverse subjects, and watched Mekeo society change profoundly. In this period Mekeo chieftainship has disappeared and people now claim that they live a modern life based on money income from selling betel to the ever-expanding market in Port Moresby.

Conclusion

This book has sought to answer the question of what caused the final transition into *Homo sapiens*, and to explain "What makes us human?"(Pasternak 2007). That transition must have been triggered by a new adaptive practice that early humans engaged in to survive. One such new practice might very well have been some form of trade, as argued by Ofek (2001). But early humans probably did not invent trade wholesale from day one. Trade must initially have been part of a more generalized form of exchange that had the potential to *become* trade. Our early ancestors must have found a new way of sharing resources between groups, and this could only be done by establishing some kind of relationship between the groups that mediated this sharing. Early human groups had to be connected somehow in order to be able to share, and the only thing that could have connected them was kinship. Kinship enabled early humans to forge alliances through which resources could travel. This would eventually lead to some form of barter and then to trade. I therefore argue that kinship was the innovative practice that caused speciation into modern humans.

This speciation associated with kinship and resource sharing probably did not happen before the last mega ice age, but rather during the ice age. In the favorable conditions that preceded the ice age, the existing population would merely have proliferated into parallel *Homo ergaster* groups that the environment could still sustain. But the harsh conditions that prevailed during the ice age would have isolated some groups in small pockets where they would have to find new means of survival, leading to fast speciation. But they did not just turn into businessmen overnight. They had to mobilize what they had right there and then: namely, their social relations.

This scenario has not been proposed by other scholars of human evolution, who have tended to overlook the fact that human evolution is based on adaptation to the environment, seeing it instead as the result of internal human group processes, qualities, and interactions peculiar to the human lineage. Most of these explanations elaborate on a kind of standard explanation that has even found its way into anthropological textbooks (Keesing 1981). In these accounts, human evolution was almost unproblematic, even ineluctable. Given the social nature of the hominine lineage, evolution into *sapiens* was bound to happen sooner or later.

These explanations favor humans as the winners of the evolutionary race, and attribute this to the fact that they began to make tools very early on, which involved the capacity to plan a sequence of steps (although many other animals also use tools of various kinds); they hunted, which required the capacity to communicate (alas, lions and other species also hunt in groups); and they lived in campsites that required cooperation (but then again so do lions and monkeys, *inter alia*). Together, so the story goes, these practices would ultimately produce intelligence and culture among humans in all kinds of environments: Europe, Java, China. At least, nowadays, most scientists have long since given up on the tool narrative, since it is now clear that new human species typically used their predecessors' tools for a very long time before deciding it was time to invent new ones.

Other scholars have tried to identify a defining moment or a new crucial skill that shaped the steady march toward intelligence and culture. They have searched for a key moment in the evolution of humans that was all important in defining the human condition and which would have changed human life and determined all subsequent developments.

There are several main contenders in this search. Burton (2009) proposes that the domestication of fire was the turning point in human evolution. He believes this gave premodern humans a fireplace or hearth that would greatly have enhanced sociability between the group members. He also thinks that even the acquisition of language was enhanced by the extended day that fire afforded (Burton 2009: 11). And it is certainly undeniable, as Pyne observes, that fire played a prominent role in human societies. All human societies have made use of fire, and all have altered the fire regimes of the lands they encountered (Pyne 2001: 27). As he boldly states: "we are fire creatures from an ice age" (Pyne 2001: 30). There is no doubt that fire has had a significant influence on human lives throughout the ages, and that fire made it possible to live in very diverse environmental settings.

However, fire seems to have been with the human lineage ever since *Homo erectus*, who did not make any progress at all for a very long time. So it is a little far out to suggest that sitting around a campfire for long enough would eventually have gotten humans talking, and even more so to suggest that this habit mysteriously led to speciation. It certainly did not for the *Neanderthals*. They just sat around their fires for a great number of years and then became extinct. Still, fire is important to human communities, past and present. From the earliest days, fire has helped humans to manage the environment, it provided them with light and heat, and it made it possible for early humans to engage in another all-important and universal practice, that of cooking.

According to Wrangham, it was cooking that made us human (2009). He observes that cooking increases the value of food, and that this changed the human body. Not only that, he even finds that cooking changed the human brain, our conception of time, and our social way of life. Cooking turned humans into consumers of external energy "and thereby [it] created an organism with a new relationship to nature, dependent on fuel" (Wrangham 2009: 2).

But cooking also arrived on the scene with *Homo erectus* (ergaster) and was closely connected to hunting. *Ergaster* hunted large animals but their teeth are known to have grown smaller over time, suggesting that they had learned to cook their game (Wrangham 2009: 8). Wrangham also finds that the arrival of *Homo erectus* marked the genesis of contemporary humans, and that from then on it was only a matter of time and brain growth before modern humans emerged some two hundred thousand years ago (Wrangham 2009: 5). Wrangham seems to think that cooking ushered humans across a crucial threshold and onto a direct path toward intelligence and culture: no further adaptive changes were required, just evolution working its incremental magic. But eating cannot be the reason for speciation, otherwise all well-nourished animals would have evolved in a similar way to humans. Cooking food is more likely to have led to obesity than speciation.

Ofek (2001) is yet another contender within this approach to evolution. He considers trade to be the most distinctive human practice, and concludes that it must have been this that propelled human evolution forward. According to Ofek, trade provided humans with a "Second Nature" (Ofek 2001). As Ofek rightfully observes, "[c]ontemporary human societies literally live by exchange and die by exchange" (2001: 23). So trade must have played a vital role in human evolution.

Trade becomes a "second nature" because with it human subsistence is no longer conditioned by local resources, and at the same time the availability of a single resource can spur migration to the most isolated locations as long as there is a demand for that particular resource in the system. And the migrating populations are able to live in such isolated places because they can obtain provisions through long-distance trade (Ofek 2001: 178).

Trade relocates resources and relieves populations from the constraints of the immediate environment. This is surely an adaptive change with major implications for the spread of modern humans to all corners of the world. Ofek finds that trade must have been "invented" when *Homo* began to hunt large game and use fire. This led to a situation where humans had to cook their meat and therefore were in dire need of a steady fire, so groups would have specialized either in hunting or making fire, and would meet to exchange game for fire (Ofek 2001: 143–52). Thereby he argues that trade preceded the emergence of modern humans, eventually producing them by means of a general evolutionary push. But in fact, all the paraphernalia associated with trade is associated with a much later period, after the last major ice age.

Unlike all other animal communities that wax, wane, and become extinct as their environment changes, in the above accounts the human lineage is strangely disconnected from its environment. Human dynamism seems to come from within or from among the group's own members. Humans' innate skills and organizational forms are thought to have propelled them forward. Certainly not adaptation. This may be due, in some part, to some kind of anthropocentric self-glorification. Still, why should we believe that humans were special from the start? That we were the only creatures miraculously exempt from the forces of adaptation, apart from our initial venture onto the savannah in response to a drying climate?

Genetics also muddy this otherwise appealing story. Genetic studies have shown that all modern humans descended from a very small population somewhere in Africa that managed to survive the last major ice age some 200,000 year ago. We now know that some groups went through a relatively rapid speciation at a time when they had already been using fire and cooking for a long time. The period when that speciation occurred must be the true turning point in human evolution, the one that produced *Homo sapiens*.

We also know that all human species evolved in Africa, and from here they spread out in different waves around the globe. Human evolution was therefore clearly not a worldwide process or part of a general evolutionary thrust that propelled only humans, out of all other species, forward toward their unique intelligence. It was not the *ergaster*, *erectus*, or *Neanderthal* populations that slowly turned into modern humans by means of fire, cooking, or trade.

All these phenomena are important ingredients in the present human condition and all are universally practiced by modern humans. The problem is that they had all already been adopted by the prehuman population, according to the above mentioned authors. As Tattersall writes, "[w]hatever the exact case, what is evident is that the human capacity was not predicted by anything that preceded it. It does not represent an incremental improvement upon what went before, but rather something truly new" (2007: 131–45).

AN EVOLUTIONARY SCENARIO FOR *HOMO SAPIENS*

To identify the "truly new" element in human evolution, we have to come up with a scenario that can explain where, and how, the environment encouraged speciation. During the last ice age (which lasted from 220 to 130 kya), much of Africa dried up leaving only small pockets of isolated *ergasters* around the continent. If these small groups had not been isolated there would have been no foundation for speciation. There is no evidence of an African multiregional evolutionary scenario in which every isolated group of *ergasters* turned into *Homo sapiens*.

Not every survival pocket would have posed the same practical problems, and this would have given the different groups different evolutionary incentives. We could, in principle, have seen the emergence of different kinds of new *Homo something*, whose characteristics would have depended on the environmental challenges in the different local pockets. This did not happen, or at least no traces have so far been found of anything like this. Most of the groups that were isolated during the ice age would have probably died out. But we know that a small group did manage to seek refuge near the coast in the Cape Floral Region that contained the Fynbos Floral Biome located in South Africa. Or, at least it is in this area where the first archeological finds from modern humans were discovered. So it would seem that not only did modern humans emerge out of Africa, they emerged specifically out of South Africa.

The prehuman ice age population took their inherited *ergaster* skills with them to South Africa. They were still able to use fire and cook their food, but as it turned out they needed a completely new skill in order to adapt to their new situation, which would have caused speciation. Without that skill, nothing would have happened and we would still all be *ergasters*. Naturally, this new skill must still be a major factor in modern human life. It must be a fundamental dimension of the present human condition, which supplements our universal use of fire and cooking. This new skill must also have greatly expanded the range of environments that humans were able to inhabit, making it possible for them to colonize all corners of the Earth. As Ofek has already suggested, this new skill must have involved an incipient form of exchange and trade.

So we know that there was a group of human ice age survivors in the Cape Floral Region. We also know that this was, and still is, a very special area. This region contains the Fynbos Biome that contains a rich source of carbohydrates in the form of tubers and bulbs. These plants were adapted to dry conditions and were therefore not particularly affected by the ice age, and therefore they offered a steady supply of resources during these harsh conditions. Running along this Biome is a coastline rich in shellfish that were not greatly perturbed by the ice age either. Together, these adjacent resource zones created a favorable environment for a small population to survive the ice age, whilst the rest of Africa suffered its consequences.

So:

1. We know from the archeological material from South Africa that different groups lived in the coastal areas and in the interior zone, and that these groups were probably not migratory between the zones (Marean 2010: 432).
2. At the same time, the ethnographic records tell us that it is very difficult to live off marine resources alone. These have to be supplemented with carbohydrates.
3. We know that females were crucial to the group's survival in these environments. Both the collection of shellfish and tubers were their domain. So it was no light matter to let one's female leave and live in another group.
4. And finally, anthropology has taught us that the most basic form of human social organization is kinship, and that kinship systems vary from place to place. But irrespective of place, the system everywhere is based on exchange relationships between groups, making kinship a means of managing resources.

If we piece these four insights together, we have the answer to human evolution. It is quite simple, and goes like this: the group living on the coast needed carbohydrates from inland, and the groups living inland needed proteins from the coast, and the only way that both groups could get what they needed was to exchange resources. This had to be done through social ties, because the exchange relationships entered into had to be lasting. With the high value of females in these conditions, groups would have tried to keep their females from leaving, or they would at least have demanded something in return for them if and when they left. So female-specific products from one group were traded for female-specific products from another. Groups would therefore strive to maintain their relationships with their daughters even after they had left to live in another group. Kinship was the result of all this. Kinship systems in all their variations are about managing resources. So the allocation of resources determines the resultant alliance network, which is expressed in a local form of kinship system.

Kinship created lasting bonds between groups through a formerly dormant matrilineal relationship between patrilineal, male-dominated groups. All that had to be done was to acknowledge the origin of the female when she moved to a new group. And *violá,* you get descent and alliances, so that everybody knew what group their females had descended from, and to whom they were related to by marriage. Otherwise there could be no consistent exchange. To ensure enduring alliances, the females became wives. And lo and behold, they unwittingly invented the primordial sign, because the wife now *represented* her group of origin, to which her new group was now related through exchange. The result of all this was culture.

LIVING IN NETWORKS

With this newfound ability to make alliances and share resources, early humans could connect even more groups with different resources and they could incorporate different specialist groups in ever expanding networks of exchange. They had become liberated from dependence on the immediate environment, allowing them, in principle, to live in territories that could not sustain them unless vital resources could be obtained through exchange. From being a matter of groups foraging in their immediate environment that had to supply all their needs, human social reproduction had turned into a matter of integration into larger networks through which resources circulated, producing a kind of "second nature," as Ofek calls it.

These networks superseded the conditions of locally constituted environments in such a way that the network became the single most important factor that informed the conditions of group life in its diverse localities. Social reality had gained an extra dimension: networks that gave local living conditions new, emergent properties. There was now more to social reality than met the eye.

From then on, social reality was shaped by forces from beyond, and even the dead played a role since they influenced the relationships of the living. Ancestors became significant in the lives of the living, enmeshing the past with the present. It could even be argued that it is the networked character of human societies that provides people with a belief in something extraordinary, an unobservable force that nonetheless influences their lives. This feeling is presumably what lies behind what, in some places, has become known as religion but which in other societies gives people the sensation of living in a double-sided world, one for the ancestors and one for the living, with dreams as a state of mind that mediates between them.

If human groups did not live in networks, their social organizations would be identically replicated even though they lived in different groups (i.e. they would exhibit *ergaster* behavior). Burroughs has also noted the significance of networks. He writes: "Networks played an essential part . . . Neanderthals . . . relied heavily on materials available locally," and he continues, "Modern humans networked well, and when times get hard they had kith and kin to run to. The *Neanderthals*, it appears, did not" (2005; Kindle Loc. 1942–43, 1944–45).

Living in networks was a game-changer. From then on, every human population had to negotiate the conditions that the network produced—its emergent properties—and since these necessarily vary from position to position in the network, groups occupying different positions within a network become differently constituted. Living in networks produces cultural differences.

Furthermore, because negotiating different emergent properties is tantamount to negotiating reality, it becomes an existential issue. Different positions in a network therefore also produced different ontologies. Culture is not just skin deep, and categories pertaining to one cultural order cannot just be superimposed on or used to understand other cultural orders. So what looks like religion or superstition from one position in a system is really just local reality seen from another position point in the system. In other words: the universal or pan-human condition is not religion per se, but the fact that one's network produces forces that have to be locally negotiated and given symbolic expression.

Such ontological differences might be accentuated by the process of construction of ethnic distinctions, identified by Barth, which takes place between two of more groups that occupy different ecological niches and exchange products between them. He finds that ethnic distinctions between groups are the foundation of all exchange systems, and that such distinctions also organize interaction between people (Barth 1969: 10). From Barth's perspective, ethnic groups are the product of categories of ascription and identification, by the actors themselves, which are used to organize interaction between people.

We thus have a two-way system that produces cultural differences: one in which a group has to negotiate the condition of its position in the network, and one in which adjacent groups ascribe characteristics to each other. In this way, even groups living in close proximity to each other can come to be culturally and socially very different from each other. Papua New Guinea is a prime example of this, containing more than 700 different language and cultural groups within a relatively small area.

Cultural differences are part and parcel of being human—and of living in networks. Even the smallest networks produce cultural differences. And because networks can exist within networks, differences can quickly proliferate. We can now begin to see why, when the early humans began to migrate out of Africa, new groups naturally became culturally different. That was part of the process of establishing new positions in a network. Staying long enough in one place would enhance that difference, both in relation to one's parent group and to the groups that migrated farther into the new territories. This is arguably the only way we can account for the immense cultural diversity we observe after new territories have been "conquered"—the Pacific, South America, and so on, given the very swift migration into these areas and the resultant great cultural diversity.

In contrast, most social scientists have treated cultural groupings as if they were singular groups living in isolation and born with inherent qualities presumed to constitute *their* unique culture. These groups might occasionally reach out to other groups in order to trade with them. But it is the system that creates differences. And having invented kinship, networks became the prime movers of social reproduction.

Networks of groups are like magical building blocks. They have unique characteristics compared to single groups foraging in isolation. Living in networks enables survival in territories that otherwise would not sustain existence. Networks therefore also create mutual dependencies, and the different groups involved have to keep things circulating. Everybody in a network has to contribute with something, since their survival depends on it. That is why networks never rest.

Social networks started with the "elementary unit of kinship." That is the minimum relationship that can connect two groups. As we saw earlier, it contains of a husband and a wife plus the uncle who gave the wife away. Or, in more abstract terms, it gives rise to an "elementary unit" of social reproduction: for example, a unit (husband-wife) featuring a link (the uncle) that himself is the product of a (husband-wife) unit. In this more abstract version we find the foundation of social reproduction in all its forms. This is the basis of the networked character of human society. Human society comprises both a cultural unit in and of itself, and links to other such units. This is what gives the system its dynamic character. A system like this can never be in equilibrium since resources can always find other ways around the networks, and the internal composition of the units is constantly changing.

Ever since the dawn of kinship, all human social reproduction has taken place in networks. In the Kongo Kingdom, as we saw earlier, these units and links were connected to the circulation of resources. Here, women and tribute moved up the system in return for bride price and valuables. This meant that the groups who were wife-"takers" had to have a resource base that they could use to "pay" for these socially important valuables. But at the same time, those occupying the highest positions had difficulties getting their daughters married since there were no groups of the same social standing who could pay them bride price. Therefore these daughters married slaves, who were positioned outside the kinship system. Another option in a tricky situation like this is to marry one's daughters into highly ranked groups in other similar systems so that the daughters, so to speak, marry in circles among differently located upper classes, such as kings—as historically happened, for instance, among the European royalty.

So, what starts out as an alliance based on kinship between two or more groups can very fast be *divided* into subnetworks, *expanded* to incorporate other networks, or *connected* to other networks, greatly enhancing the total circulation of resources among groups. Connecting or subdividing networks also opens up for the possibility of creating local hierarchies which can then grow if they are also able to connect to the multiple other networks through which different resources circulate.

At times, networks may cluster in one area, especially if they are able to connect to several other and larger networks through which vital resources circulate. But these are highly unstable structures in the long run because resources can always find new paths through which to circumvent the systems. At the same time, there will always be some positions in networks that have precious little to offer, and which are mostly left to rely on local resources. So networks of reproduction will tend to oscillate between *clusters* and *units*. This is the civilizing process, where clusters become the centers of civilizations and units become supply zones of different kinds of raw material (see, e.g., Friedman 1994).

But the networked character of cultural differences does not only pertain to whole populations in interaction. It can be found at all levels in society. Such processes of cultural differentiation can even be found in everyday workplaces where different networks also interact, for example, in an *institution* one find departments, companies, and countries interacting with *individual* roles, tasks, positions and backgrounds, producing networks that interact and creates different "realities" and cultures in each position or section of a workplace.

EVOLUTION AND THE SOCIAL SCIENCES

We are now in a position to explain evolution. Due to resource scarcity, the evolving humans invented kinship to integrate the resources from complementary ecological zones. In the process, females became wives who represented their groups of origin, introducing the symbolic, and subsequently culture. Kinship consolidates networks between groups. Networks produce different emergent properties in their different positions. Emergent properties are negotiated by the groups living there, producing different forms of cultural organization.

We thus have the two characteristics of human society that the social sciences have grappled with for so long, erroneously separating them into different social spheres: the economy and culture, each with its own science. But these phenomena are both interdependent aspects of human life in networks of reproduction. It is their interconnections that should be the point of departure for the social sciences.

It seems that anthropology has taken cultural relativism too far. As Wells argues, from a genetic viewpoint "we need a new model to explain cultural diversity and thus we need to move beyond cultural relativism," and he continues: "anthropology faces a major challenge, cultural relativism can no longer be taken as the point of departure" (2011: Kindle Loc. 1136 and 1138).

Cultural relativism was a noble project in its time. It arose in response to a contemporary trend to biologize human behavior which emphasized genetic classifications and their presumed mental correlates that could be used to justify restrictive immigration policies and racism. But it also turned culture into a symbolic prison, as expressed by the influential anthropologist Leslie White: "Culture may thus be considered as a self-contained, self-determined process; one that can be explained in terms of itself" (White 1949).

As it turns out, this was more of a political agenda than a scientific one. Genetics shows that we are all descendants from a very small group in Africa, making racism a purely ideological construct. Cultural relativism took that which needs to be explained as a given, namely cultural diversity. Under this scientific regime, cultural variation remains an unresolved mystery. Kinship was the great transformation in human evolution. Kinship is what makes us human. It enables networks that connect resources and produce cultural differences: the two fundamentals of the human condition. There is no such thing as a cultural group per se, there are only positions in a network that have to be negotiated, and conditions change continually in every position, as do the cultural categories that pertain to each position. This, of course, produces a form of cultural relativism since cultural orders have different ontologies, but these differences should be seen in relation to the networks involved in the reproduction of the involved cultural units. Hopefully, there will be no more anthropological studies of "a people," since peoples have no separate existence or inherent qualities. Gone are the days of Franz Boaz and cultural relativism.

Cultural relativism, substantivist economics, and formalist economics err in looking solely at the involved units to the complete exclusion of the networks that integrate the units and make them different from each other. The debate between these positions essentially assumes that people act in isolated social units, asking if individuals within those units are able to transform their society or if they are merely a product of it. In the last case, culture is seen as the determining element in human life; in the first case people seemingly operate in an empty space only guided by their self interest in the pursuit of scarce resources.

All these theories conceptualize culture as though it consisted of integrated and closed units separated from each other. These units either operate according to their own internal laws that are specific to each and anyone, or according to some supposed universal drive found within all humans, which creates cultural differences out of nowhere. Both these views are based on an optical illusion that results from looking only at the units and then claiming that the specificity of each unit must be found within it. This view predates our present knowledge of the colonization of the Americas and the Pacific. Now, cultural relativism seems hard to reconcile with a relatively swift colonization and its resultant cultural variation.

We know from evolution that the great transition in the human population resulted from a shift from foraging in isolated groups that had to obtain all the resources they needed within their own territories, to integration in networks that allocated resources among groups. Consequently, the group has no life of its own, and therefore the dichotomy between materialists and relativists, and between methodological individualism and methodological holism, is irrelevant.

From an evolutionary point of view, what we are probably dealing with are actually quite free individuals who were able to effectively negotiate their local conditions. At the same time, these conditions were not only produced locally, meaning that the early humans were constrained by conditions that they did not produce and which they could not escape. There are only a limited number of strategies available in any given position, at any given time in a network. This situation tends to produce uniformity in the options available locally, which informs the character of social relations in any given position in the network. This leaves us with individuals and networks as complementary aspects of social reproduction.

What is more, the assumption that human self-interest is the driving force of "the market" seems unlikely in evolutionary terms. It seems more likely that when early humans increasingly found themselves in ecological zones that did not supply them with the resources they needed, they started to share between the zones. They became social. This should make us deeply skeptical of economic theories about self-regulating markets that will efficiently allocate resources among all people. Networks do not have any inbuilt equilibrium or any inherent means of distributing resources equitably. On the contrary, there will always be the possibility for endless accumulation in some positions in a network.

It follows that it is difficult to single out a special sphere in this social organization and label it "the economy" and then claim that it has a life of its own or possesses a separate dynamic—or even worse, that it determines the character of the social. At the root of human social organization is the social organization of resources. In the beginning, this was based on kinship which made it possible to share resources between groups. So, even the basic assumptions of economics seem questionable. The economy is not driven by selfish individuals in a situation of limited resources. Rather, it is driven by sharing—the basic trait of kinship and humanity in general.

So it seems that the act of sharing is the *primus motor* of human society. Sharing is the primordial act of being human. But sharing is not a great equalizer that works for the benefit of all. Sharing creates distinctions because it takes place in networks with unequal opportunities. Sharing in a network is a complicated matter. People or groups put something out there in the network in the hope of getting something back. Then, of course, there is the risk that your "products" fall into disfavor, are no longer in demand, or are out-competed by other sources. Sharing also creates dependencies, and therefore it has the capacity to create social hierarchies and power relations.

Depending on the scale and number of networks involved, sharing will at one point in history, or in some position in the network, appear in the form of trade because its personalized character disappears from view. This is what has been elevated to a form of a general economic force that works on its own, but then one is missing the point.

Furthermore, the process of human evolution teaches us that it was not tinkering with tools that brought the human lineage forward. And consequently, all derivative theories that link social dynamics to some form of material basis, mode of production or technological level must be rethought. Human development was not an unstoppable technological journey toward intelligence and cultural complexity. Inventions always came after a social reorganization of the human population.

Another consequence of looking at the social sciences from an evolutionary point of view is that there can be no real science unless the emergent properties produced in the network "position" of the scientific community are called into question in scientific investigations. Otherwise, most of what is being investigated merely constitutes the emergent properties of that particular position in the larger network.

Alternatively, these emergent properties can be investigated together with how they are negotiated locally and how this produces a local reality, instead of just using categories in one part of the system to characterize all forms of human organizations, as has, to a large extent, happened with the study of "religion." We need a global science that incorporates all, or none, of the emergent properties of the network. This could be the agenda of a new global, evolutionary anthropology. The West is not necessarily the best, after all.

Emergent properties have given rise to a host of scientific concepts and theories. They have left us with cultural relativism, and they have equipped individuals with endless supposed needs and scarce resources. They have led many to believe that we can only interact with the environment through our cultural concepts, even though these same concepts result from the negotiation of reality as seen from the different positions of a reproductive network. This is but a scientific hall of mirrors.

Living in networks would also mean that humanity is not on the march toward total intelligence, in which some groups are far more advanced than others. So-called science is a recurrent historical and geographical phenomenon. Science comes and goes in different positions in different networks. At one time, the Greeks discovered that the Earth was round, and they "invented" democracy (see, e.g., Friedman 1994). Chinese philosophers seem to have invented all kinds of technological improvements and also a medical system that seems to work, but which is completely different from Western medicine.

Intelligence, thus, seems to be linked to the character of the position in the networks. Intelligence is not a general evolutionary process working throughout the universe. It would be a complete marvel if the same evolutionary process had taken place anywhere else. There seems to be no general evolutionary thrust that drives all creatures everywhere in the universe toward intelligence and culture. Maybe we are alone in the universe.

So humans never left nature. Rather, we did exactly what adaptation prescribed—like any other animal—but we called it culture. Humans do what humans do, they make cultural distinctions and this is what enables them to survive. If there is a next level of human cultural complexity, it will involve humans coming to terms with their evolutionary origins and making them work for them—acknowledging the networked premises of their social organization and the insights this carries: notably that we are locked in space, and limited in perception due to our inescapable positioning in networks.

Maybe we can transcend this predicament, but then the first step would be to acknowledge this as a basic premise of all knowledge. We need a global science that incorporates our evolutionary origins and our present cultural diversity—or a science that can transcend the multitude of ontologies that living in networks has conferred upon the human population. If a social theory cannot explain, or be applied to our evolutionary origins, it is probably wrong. We are not as free from adaptation as we think.

Bibliography

Allen, J. "Fishing for Wallabies: Trade as a Mechanism for Social Interaction, Integration and Elaboration on the Central Papuan Coast." In *The Evolution of Social Systems*, edited by J. Friedman and M. J. Rowlands, 419–55. London: Duckworth, 1977.

Allen, N. J. *Categories and Classifications: Maussian Reflections on the Social, Methodology and History in Anthropology*, volume 8. New York: Berghahn Books, 2000.

Appadurai, A. *Modernity at Large: Cultural Dimensions of Globalization*, vol. 1. Minneapolis: University of Minnesota Press, 1996.

Asad, T. "The Construction of Religion as an Anthropological Category." In *A Reader in the Anthropology of Religion*, edited by M. Lambek, 114–32. Oxford: Blackwell Publishing, 2002.

Balme, J., I. Davidson, J. McDonald, N. Stern, and P. Veth. "Symbolic Behaviour and the Peopling of the Southern Arc Route to Australia. *Quaternary International* 202, no. 1–2 (2009): 59–68.

Barnard, A. *Social Anthropology and Human Origins*. Cambridge, UK; New York: Cambridge University Press, 2011.

———.*Genesis of Symbolic Thought*. Cambridge, UK; New York: Cambridge University Press, 2012.

Barth, F. (Ed.) *Ethnic Groups and Boundaries: The Social Organization of Culture Difference*. (Results of a symposium held at the University of Bergen, 23rd to 26th February 1967). Universitetsforlaget.

Barth, F. *Ethnic Groups and Boundaries: The Social Organization of Culture Difference*. Long Grove, IL: Waveland Press, 1998.

———."Toward a Richer Description and Analysis of Cultural Phenomena." In *Anthropology beyond Culture*, edited by R. Fox and B. King, 23–36. New York: Bloomsbury Academic, 2002.

Beinhocker, E. D. *The Origin of Wealth: Evolution, Complexity, and the Radical Remaking of Economics*. Boston: Harvard Business School Press, 2006.

Bergendorff, Steen. "The Reproduction of the Mekeo Chieftainship: The Complexity of Cultural Contact." *Folk: Journal of the Danish Ethnographic Society* 35 (1993):37–64.

———. "Literal Meanings: The Case of Mekeo Sorcery." *The Asia Pacific Journal of Anthropology* 6, no. 1 (2005): 57–79.

———. *Faingu City: A Modern Mekeo Clan in Papua New Guinea, Lund Monographs in Social Anthropology*. Lund, Sweden; Bromley, Kent [England]: Lund University Press; Chartwell-Bratt, 1996.

———. "The Mekeo Sorcerer and His Magic." *The Asia Pacific Journal of Anthropology* 4, no. 1 (2003): 72–87.

———. "Semantic Frameworks: Methodological Reflections on How to Understand Mekeo 'Sorcery' and 'Magic.'" *Field Methods* 19, no. 4 (2007): 407–24.

———. "Cultural Complexity and Religion: Persistent Interaction and Conceptual Order in Mekeo Society, Papua New Guinea. *Culture & Religion* 9, no. 2 (2008): 109–24.

———."Contextual Translation—Concepts and Practices among the Mekeo of Papua New Guinea." *Oceania* 79, no. 3 (2009): 225–37.

———.*Simple Lives, Cultural Complexity: Rethinking Culture in Terms of Complexity Theory.* Lanham, MD: Lexington Books, 2009.

Bickerton, D. 2009. *Adam's Tongue: How Humans Made Language, How Language Made Humans.* 1st ed. New York: Hill and Wang, 2009.

Borofsky, R. (ed.). *Assessing Cultural Anthropology.* New York: McGraw-Hill, 1994.

Bramble, D. M., and D. E. Lieberman. "Endurance Running and the Evolution of Homo." *Nature* 432, no. 7015 (2004): 345–52.

Burling, R. *The talking Ape: How Language Evolved, Studies in the Evolution of Language 5.* Oxford; New York: Oxford University Press, 2005.

Burroughs, W. J. *Climate Change in Prehistory: The End of the Reign of Chaos.* Cambridge; New York: Cambridge University Press, 2005.

Burton, F. D. 2009. *Fire: The Spark that Ignited Human Evolution.* Albuquerque: University of New Mexico Press, 2009.

Cann, R. L., and A. C. Wilson. "The Recent African Genesis of Humans." *Scientific American Special Edition* 13, no. 2 (2003): 54–61.

Carrithers, M. "Why Humans Have Cultures." *Man* 25, no. 2 (1990): 189–206.

Cela-Conde, C. J., and F. J. Ayala. *Human Evolution: Trails from the Past, Oxford Biology.* Oxford; New York: Oxford University Press, 2007.

Chapais, B. *Primeval Kinship: How Pair-Bonding Gave Birth to Human Society.* Cambridge, MA: Harvard University Press, 2008.

Clifford, J. *The Predicament of Culture.* Cambridge, MA: Harvard University Press, 1998.

Cochran, G., and H. Harpending. *The 10,000 Year Explosion: How Civilization Accelerated Human Evolution.* New York: Basic Books, 2009.

Committee on the Earth System Context for Hominin, Evolution, and Council National Research. *Understanding Climate's Influence on Human Evolution.* Washington, DC: National Academies Press, 2010.

Coolidge, F. L., and T. G. Wynn. *The Rise of Homo Sapiens: The Evolution of Modern Thinking.* Chichester, UK; Malden, MA: Wiley-Blackwell, 2009.

Danida, *Denmark's Development Policy—Strategy.* Copenhagen: Ministry of Foreign Affairs, 2009.

Dunbar, R. I. M. *The Human Story: A New History of Mankind's Evolution.* London: Faber and Faber, 2004.

Dunbar, R. I. M., C. Knight, and C. Power. *The Evolution of Culture: An Interdisciplinary View.* New Brunswick, NJ: Rutgers University Press, 1999.

Egidi, P. V. M. "Le Leggi e le cerimonie del matrimonio nella tribu di Mekeo." *Rivista di Antropolgia* 17, no.1–2 (1912): 217–29.

Elkin, H. "The Northern Arapaho of Wyoming." PhD, Dissertation. Appleton-Century Co., Columbia University, New York, 1940.

Elton, S. "The Environmental Context of Human Evolutionary History in Eurasia and Africa. *Journal of Anatomy* 212, no. 4 (2008): 377–93.

Evans-Pritchard, E. E. *Theories of Primitive Religion, Sir D. Owen Evans Lectures 1962.* Oxford: Clarendon Press, 1965.

Fox, R. The Harvey Lecture Series. The Evolution of Mind: An Anthropological Approach. *Journal of Anthropological Research* (1979): 138–56.

Friedman, J. *Cultural Identity and Global Process, Theory, Culture & Society.* London: Sage, 1994.

Friedman, K. E. "External Exchange and the Transformation of Central African Social Systems." In *Historical Transformations: The Anthropology of Global Systems*, edited by K. E. Friedman and J. Friedman, 231. Lanham, MD: AltiMira Press, 2008.

Gamble, C. *Origins and Revolutions: Human Identity in Earliest Prehistory*. New York: Cambridge University Press, 2007.
Geertz, C.. *The Interpretation of Cultures: Selected Essays*. New York: Basic Books, 1973.
Gibson, Kathleen R. General Introduction: Animal Minds, Human Minds in Gibson, K. R., and T. Ingold. *Tools, Language, and Cognition in Human Evolution*. Cambridge, England; New York: Cambridge University Press, 1993.
Hall, S., and Open University. *Representation: Cultural Representations and Signifying Practices, Culture, Media, and Identities*. London; Thousand Oaks, CA: Sage in association with the Open University, 1997.
Harding, T. G. "Precolonial New Guinea Trade." *Ethnology* 33, no. 2 (1994): 101.
Hart, D., and R. W. Sussman. *Man the Hunted: Primates, Predators, and Human Evolution*. New York: Westview Press, 2005.
Henshilwood, C. S., and B. Dubreuil. "The Still Bay and Howiesons Poort, 77–59 ka." *Current Anthropology* 52, no. 3 (2011): 361–400.
Herskovits, M. J. *The Economic Life of Primitive Peoples*. New York, London: A.A. Knopf, 1940.
Jones, M. *Feast: Why humans Share Food*. Oxford; New York: Oxford University Press, 2007.
Keesing, R. M. *Cultural Anthropology: A Contemporary Perspective*. 2nd ed. New York: Holt, Rinehart, and Winston, 1981.
Keesing, R. M., and A. Strathern. *Cultural Anthropology: A Contemporary Perspective*. 3rd ed. Fort Worth: Harcourt Brace College Publishers, 1998.
Kenneally, C. *The First Word: The Search for the Origins of Language*. New York: Viking, 2007.
Klein, R. G., and B. Edgar. *The Dawn of Human Culture*. New York: Wiley, 2002.
Kowald, C. "Mekeo Habits and Customs of the Mekeo District (Central Division)." *British New Guinea Annual Report*. Appendix S. 61–65. 1893.
Lakoff, G., and M. Johnson. *Metaphors We Live By*. Chicago: University of Chicago Press, 1980.
———.*Philosophy in the Flesh: The Embodied Mind and Its Challenge to Western Thought*. New York: Basic Books, 1999.
Leach, E. R. *Rethinking Anthropology, Monographs on Social Anthropology*. London: Athlone Press, 1961.
Leonard, W. R. Food for Thought. *Scientific American Special Edition* 13, no. 2 (2003): 62–71.
Levi-Strauss, C. "The Structural Study of Myth." In *Structural Anthropology*. Suffolk: Penguin Press, 1963.
———. *Structural Anthropology. Vol. 1*. Hammondsworth, NY: Penguin Books, 1968.
———. *The Elementary Structures of Kinship*. Rev. ed. Boston: Beacon Press, 1969.
———. *The Raw and the Cooked*. New York: Octagon Books, 1979.
Lewin, R., and R. Foley. *Principles of Human Evolution*. 2nd ed. Malden, MA: Blackwell Publishing Co., 2004.
MacGregor, W. Dispatch reporting visit of inspection to the St. Joseph River district. British New Guinea Annual Report 60, 76–83.
Mann, N. "Dietary Lean Red Meat and Human Evolution. *European Journal of Nutrition* 39, no. 2 (2000): 71.
March, J. G. and J. P. Olson. The logic of appropriateness. Arena: Centre for European Studies, University of Oslo.
Marean, C. W. "Pinnacle Point Cave 13B (Western Cape Province, South Africa) in context: The Cape Floral Kingdom, Shellfish, and Modern Human Origins." *Journal of Human Evolution* 59, no. 3–4: (2010) 425–43.
———. "When the Sea Saved Humanity." *Scientific American* 303, no. 2: (2010) 54–61.
Mithen, S. J. *The Prehistory of the Mind: A Search for the Origins of Art, Religion, and Science*. London: Thames and Hudson, 1996.
Momin, A. R. "Philosophical Anthropology in the Discourse of Sufism." Alternative Models in Anthropology, Indira Gandhi National Centre for the Arts, New Delhi, 2001.
Morris, B. *Anthropological Studies of Religion: An Introductory Text*. Cambridge University Press, 1987.

Nettle, D., and R. I. M. Dunbar. "Social Markers and the Evolution of Reciprocal Exchange." *Current Anthropology* 38 (1997): 93–99.

Ofek, H. *Second Nature: Economic Origins of Human Evolution*. Cambridge, UK; New York: Cambridge University Press, 2001.

Opie, K. and C. Power. "Grandmothering and Female Coalitions: A Basis for Matrilineal Priority?" In Callan, H., Dunbar, R., and James, W. *Early Human Kinship*. Oxford: Blackwell, 2008.

Oppenheimer, S. *Out of Eden: The Peopling of the World*. London: Constable & Robinson, 2003.

Pagel, M. D. *Wired for Culture: Origins of the Human Social Mind*. 1st ed. New York: W. W. Norton, 2012.

Parkington, J. *Coastal settlement between the mouths of the Berg and Olifant Rivers, Cape Province*. S. African Archaeological Bulletin 32 (1976): 150–157

Pasternak, C. A. *What Makes Us Human?* Oxford: Oneworld, 2007.

Polanyi, K. *The Great Transformation*. New York; Toronto: Farrar & Rinehart, 1944.

Pyne, S. J. *Fire: A Brief History, Cycle of Fire*. Seattle: University of Washington Press, 2001.

Radcliffe-Brown, A. R. *The Social Organization of Australian Tribes, The "Oceania" Monographs, no. l*. Melbourne: Macmillan & Co., limited, 1931.

Radcliffe-Brown, A. R., and A. Kuper. *The Social Anthropology of Radcliffe-Brown*. London; Boston: Routledge & Kegan Paul, 1977.

Rappaport, R. A. *Ritual and Religion in the Making of Humanity, Cambridge Studies in Social and Cultural Anthropology 110*. Cambridge; New York: Cambridge University Press, 1999.

Ribeiro, G. L., and Escobar, A. *World Anthropologies: Disciplinary Transformations Within Systems of Power*. New York: Berg Publishers, 2006.

Saint-Paul, G. On market forces and human evolution. *Journal of Theoretical Biology, 247*(3), 397–412.

Saraswati, B. "Alternative Paradigms in Anthropology: The Cosmic Anthropological Principle." *The Nature of Man and Culture: Alternative Paradigms in Anthropology*. New Delhi: Aryan Books International, 2001.

Schneider, H. K. *Economic Man; The Anthropology of Economics*. New York: Free Press, 1974.

Seabright, P. *The Company of Srangers: A Natural History of Economic Life*. Princeton, NJ; Oxford: Princeton University Press, 2004.

Seligman, C. G., F. R. Barton, and E. L. Giblin. *The Melanesians of British New Guinea*. Cambridge: Cambridge University Press, 1910.

Skorupski, J. *Symbol and Theory*. Cambridge: Cambridge University Press Archive, 1976.

Steward, J. H. *The Economic and Social Basis of Primitive Bands*. Chicago: Bobbs-Merrill, 1936.

Stone, L., P. F. Lurquin, and L. L. Cavalli-Sforza. *Genes, Culture, and Human Evolution: A Synthesis*. Malden, MA: Blackwell, 2007.

Strathern, M. *The Gender of the Gift: Problems with Women and Problems with Society in Melanesia*, vol. 6. Oakland: University of California Press, 1988.

Tambiah, S. J. *Magic, Science and Religion and the Scope of Rationality*. Cambridge: Cambridge University Press, 1990.

Tattersall, I. "Human Evolution and the Human Condition." In *What Makes Us Human*, edited by C. Pasternak, 133–45. London: Oneworld Publications, 2007.

———.*Masters of the Planet: Seeking the Origins of Human Singularity*. New York: Palgrave Macmillan, 2012.

Taylor, J. R. *Linguistic Categorization*. Oxford: Oxford University Press, 1989.

Trouillot, M. R. *Anthropology and the Savage Slot: The Poetics and Politics of Otherness*. In R. G. Fox (ed.) Recapturing anthropology: working in the present (17–44). Santa Fe, NM: School of American Research Press, 1991.

———."Adieu, Culture: A New Duty Arises. In *Anthropology beyond Culture*, edited by R. Fox and B. King, 37–60. New York: Bloomsbury Academic, 2002.

Tucker, V. "'Introduction': A Cultural Perspective on Development." In *Cultural Perspectives on Development*, edited by V. Tucker, 1–21. London: Routledge, 1997.

Vigilante, T. "Analysis of Explorers' Records of Aboriginal Landscape Burning in the Kimberley Region of Western Australia." *Australian Geographical Studies* 39, no. 2 (2001): 135.

Wade, N. *Before the Dawn: Recovering the Lost History of Our Ancestors*. New York: Penguin Press, 2006.

Wade, N. *The Faith Instinct: How Religion Evolved and Why It Endures*. New York: Penguin Press, 2009.

Wells, S. *The Journey of Man: A Genetic Odyssey*. New York: Random House Trade Paperbacks, 2002.

———. *Pandora's Seed: The Unforeseen Cost of Civilization*. New York: Random House, 2010.

———. *Deep Ancestry: How DNA Reveals the Roots of Your Family Tree*. 2nd ed. Washington, DC: National Geographic, 2011.

Wheeler, W. 2006. *The Whole Creature: Complexity, Biosemiotics and the Evolution of Culture*. London: Lawrence & Wishart, 2006.

White, L. A. *The Science of Culture, a Study of Man and Civilization*. New York: Farrar, Straus, 1949.

Will, M., J. E. Parkington, A. W. Kandel, and N. J. Conard. "Coastal Adaptations and the Middle Stone Age Lithic Assemblages from Hoedjiespunt 1 in the Western Cape, South Africa." *Journal of Human Evolution* 64, no. 6 (2013): 518–37.

Wilson, E. O. *The Social Conquest of Earth*. 1st ed. New York: Liveright Publishing Corporation, 2012.

Wong, K. "The Morning of the Modern Mind." *Scientific American* 292, no. 6 (2005): 86–95.

Wrangham, R. W. *Catching Fire: How Cooking Made Us Human*. New York: Basic Books, 2009.

Yengoyan, A. A. "Economy, Society, and Myth in Aboriginal Australia." *Annual Review of Anthropology* 8 (1979): 393–415.

Index

About the Author

Steen Bergendorff has a PhD in anthropology from the University of Copenhagen, Denmark, and a doctorate from Lunds University, Sweden. He has been assistant and associate professor at International Development Studies, Roskilde University, Denmark, since 1997, and is now emeritus. His research interests are in historical anthropology and world-system theory, with an offshoot to a study in understanding culture as a complex dynamic system, which has now turned into an interest in human evolution. Some of his publications include *Faingu City—A Modern Mekeo Clan*, "Cultural Complexity and Development Policy" (*European Journal of Development Research*, 2007), *Simple Lives; Cultural Complexity—Rethinking Culture in Terms of Complexity Theory* and "Contextual Translations—Concepts and Practices among the Mekeo of Papua New Guinea" (2009).

Lightning Source UK Ltd.
Milton Keynes UK
UKOW02n1213010316

269398UK00001B/29/P

9 781498 524179